# BENJAMIN FRANKLIN ISHERWOOD NAVAL ENGINEER

THE YEARS AS ENGINEER IN CHIEF, 1861-1869

BY

EDWARD WILLIAM SLOAN, III

UNITED STATES NAVAL INSTITUTE
*Annapolis, Maryland*

This book has been brought to publication by the generous assistance of Marguerite and Gerry Lenfest.

First Naval Institute Press paperback edition 2012

ISBN 978-1-59114-793-0

*Copyright © 1965*
*By United States Naval Institute*
*Annapolis, Maryland*

*Library of Congress Catalogue Card No. 65-22011*

*Printed in the United States of America*

*To my son, Michael*

# Preface

The shot-and-shell approach to the writing of naval history has, for years, unnecessarily limited our knowledge and appreciation of the United States Navy's role during time of war. Shunted aside by the romance of the broadside, naval technology and, in particular, naval administration have at best received brief mention. Consequently, those men who served behind the lines in wartime have earned scant historical notice, regardless of their immediate or ultimate contributions to the service and to their nation. Benjamin Franklin Isherwood is one such man.

Isherwood's naval career, spanning those forty years which saw the transition from the first steam warships of the 1840's to the birth of the modern American Navy in the 1880's, reached its zenith during his eight-year service, from 1861 to 1869, as engineer in chief of the Navy. During and immediately after the American Civil War, he exercised his greatest influence both in the Navy and in the world of engineering. It is proper, therefore, to begin his rescue from historical obscurity by presenting him at the most active, productive, and controversial time of his life.

In no sense is this book intended to be the full biography of Benjamin Isherwood. Rather, it is a study of that relatively brief period of his life which has the greatest historical significance.

This is the story of Isherwood as a naval officer and as an engineer. These two facets of his life were inseparable, and in both areas he engaged in bitter and widespread controversy during his entire tenure as engineer in chief. His considerable contribution to the development of marine steam propulsion and to the advancement of scientific techniques in experimentation equalled in importance his administrative services to the Navy, and as an engineering pioneer of the mid-nineteenth century, his career in the realm of technology was no less arduous than that in the Navy. His efforts to carry on these two careers simultaneously and successfully during a period of national crisis best demonstrate Benjamin Isherwood's remarkable talents, resourcefulness, and dedication to duty.

With no major collection of Isherwood papers on which to base this study, I have relied on the abundant, if incomplete, official records of the Navy Department deposited in the United States National Archives. From the collected papers of Gideon Welles, Gustavus Fox, David Dixon Porter, John Ericsson, and others, I have assembled additional material, especially personal reports and observations which supplement the departmental records and shed light—and often considerable heat—on the career of Benjamin Isherwood.

Fortunately, Isherwood's granddaughter, Mrs. Madeleine Kerwin, of New York city, has been of invaluable aid in providing abundant and colorful recollections of his private life, especially for the period after his retirement from the Navy. Mrs. Kerwin spent much of her childhood living in Isherwood's house, and by the time he died she was in her thirties. My interpretation of his personality rests largely on facts and impressions which I obtained from Mrs. Kerwin in a series of interviews held during 1962. Through her detailed and perceptive recollections, I have come to know Benjamin Isherwood as a captivating and remarkable, albeit obdurate, opinionated, and thoroughly fallible human being.

I wish to thank the following people for their considered advice and generous assistance: Wilmer R. Leech and Arthur J. Breton, Manuscripts Division of The New-York Historical Society; David C. Mearns and his staff, Manuscript Division of The Library of Congress; Elbert L. Huber and his staff, Naval and Military Service Branch of The National Archives; Buford Rowland, The National Archives; P. K. Lundeberg, Howard Chapelle, Melvin Jackson, Robert Vogel, and Silvio Bedini, Smithsonian Institution; Professor Eugene S. Ferguson, Iowa State University of Science and Technology; Rear Admiral E. M. Eller, U. S. Navy (Retired), director of Naval History, Naval Historical Foundation, Department of the Navy; Joseph S. Hepburn, The Franklin Institute of the State of Pennsylvania; John Buchanan, Cornell University Library; Professor Vernon D. Tate, United States Naval Academy; Ruth M. Leonard, American Society of Naval Engineers; J. Welles Henderson, Philadelphia Maritime Museum; Herbert Lee Seward, martime consultant and professor emeritus of Mechanical and Marine Engineering, Yale University; Ruth White; Richmond D. Williams, Eleutherian Mills Historical Library, Wilmington, Delaware; C. Harold Berry, Gordon McKay, professor of Mechanical Engineering emeritus, Harvard University; Professor Ari Hoogenboom, Pennsylvania State University; the staff of the American Society of Mechanical Engineers; and Harry E. P. Meislahn, headmaster of the Albany Academy.

I would especially like to acknowledge the assistance and encouragement I have received from Rear Admiral John D. Hayes, U. S. Navy (Retired), Annapolis, Maryland. Leonard A. Swann, Jr., whose Harvard University doctoral dissertation on John Roach took him through much of the same source material and posed many of the same problems that I encountered, offered suggestions and interpretations of great value for my understanding Isherwood's role in naval affairs after the Civil War. Robert G. Albion, Gardiner professor of Oceanic History and Affairs emeritus, Harvard University, liberally contributed his extensive knowledge of naval history in all of its aspects to enrich my presentation and to clarify my perception of the naval career of Benjamin Isherwood.

*Edward W. Sloan III*

*Hartford, Connecticut*
*February, 1965*

# Contents

| | | |
|---|---|---|
| I. | *The New Engineer in Chief* | 3 |
| II. | *Building the Union Navy* | 27 |
| III. | *Isherwood and the Ironclads* | 49 |
| IV. | *The Unwelcome Pioneer* | 79 |
| V. | *The Lawyer and the Engineer* | 105 |
| VI. | *Trials and Tribulations* | 119 |
| VII. | *The Steam Bureau under Attack* | 133 |
| VIII. | *The Inventor and the Engineer* | 143 |
| IX. | *Isherwood's Masterpiece* | 159 |
| X. | *Line against Staff* | 189 |
| XI. | *Triumph of the Reactionaries* | 213 |
| XII. | *Always the Engineer* | 233 |
| | *Notes* | 245 |
| | *Bibliography* | 267 |
| | *Index* | 287 |

# *Illustrations*

*(following page 146)*

*Secretary of the Navy Gideon Welles*

*John Lenthall*

*Edward Nicoll Dickerson*

*John Ericsson*

*Gustavus Vasa Fox, assistant secretary of the Navy*

*The Sloop-of-War* Wampanoag

*General Plan of the* Wampanoag's *Boiler and Engine Rooms*

*Longitudinal View of the Main Boilers*

*Cross Sectional View of the Main Boilers*

*A Page Reproduced from* Cassier's Magazine

*Rear Admiral Benjamin Franklin Isherwood, United States Navy*

# BENJAMIN FRANKLIN ISHERWOOD
NAVAL ENGINEER

## I. *The New Engineer in Chief*

Abraham Lincoln, busy with the mass of executive nominations required of a newly inaugurated President, sent a short note, on March 22, 1861, to his Secretary of the Navy, Gideon Welles.

> Sir: I understand there is a vacancy in the office of Engineer-in-Chief of the Navy, which I shall have to fill by appointment. Will you please avail yourself of all the means in your power for determin[ing] and present me the name of [the] best man for the service. . . .[1]

As sectional tensions moved the American nation toward a crisis, no choice for an important naval office, even for one as presumably nonpolitical as engineer in chief, could be made without regard to special political pressures which now were added to the normal intraservice rivalry for such a post. Moreover, the competition for the office of engineer in chief was not confined within the service, as it would necessarily be for most naval positions.

It was entirely appropriate that a civilian marine engineer should be considered for the post, because a naval engineer in the 1860's was still, in the eyes of the Navy, a civilian in officer's dress. The difference in training and professional duties between a naval and civilian marine engineer was slight; consequently, the transition from civilian life was still easy and frequent at all levels of engineering duty. Although the naval engineer was technically a commissioned officer and his corps a recognized branch of the service since its creation in 1842, the fact remained that, in practice, he was still not a Navy man but only a glorified mechanic, especially in the eyes of line officers who sighed morosely for the days of sail.

That a leading civilian marine engineer should now have hopes for direct appointment as chief of the Navy's engineers was not unreasonable, especially as such appointments had been made several times before. Many such civilian experts saw in this office an opportunity for professional recognition and advancement; perhaps some also perceived the alluring prospect for quick wealth, derived from influence over government contracts for machinery and supplies.

The office of engineer in chief had grown in stature during the previous twenty years. Although the incumbent as yet did not command his own bureau, he still was a key figure in naval administration, working closely with his immediate superior, the chief of the Bureau of Construction, Equipment, and Repairs. Together, these men had primary responsibility for the design, construction, and maintenance of all the vessels in the United States Navy. Furthermore, in an age of rapid transition from sail to steam, the Navy depended increasingly on its engineer in chief—the man who governed that mode of propulsion which, day by day, posed a greater challenge to the "Old Navy," which for centuries had ruled the sea under clouds of canvas, undefiled by the sooty residue of man-made power.

Secretary Welles, though new in office, had anticipated the President's request, and for several days had been scrutinizing candidates for this key post in the Navy. While newspapers speculated on his ultimate choice, civilian and naval engineers alike applied what influence they could muster. Sectional considerations, so vital at this particular moment in American history, intruded to the extent that a former North Carolinian became a particular favorite because his candidacy was pressed by many leading Union men from the crucial border states, reported *The New York Times,* on March 21.

The Secretary of the Navy, however, had made up his mind. On the day following the President's request, Gideon Welles made his recommendation; and on the same day, Abraham Lincoln nominated a career naval engineer, Benjamin Franklin Isherwood, to be the new engineer in chief of the Navy.

By law, Isherwood's nomination was subject to Senate confirmation. Not for the first time, and certainly not for the last, Benjamin Isherwood met opposition as Charles Sumner, speaking for a group of civilians objecting to the nomination, rose in the Senate to present a memorial against the candidate. Referred to the Committee on Naval Affairs, this petition met a quick death as Senator John P. Hale, the chairman, reported back his committee's favorable decision on Isherwood's nomination. With the opposition squelched, the Senate, on March 27, then duly advised and agreeably consented to Benjamin Isherwood's being the new chief of the naval engineers.

In general, public reception to the appointment was cordial. Although several civilian engineers were sadly disappointed, *The New York Times* reported, the choice of Isherwood was "very gratifying to the Engineer Corps."[2] On April 13 the *Scientific American* enthusiastically endorsed Isherwood and concurred in the view of *The Times* that this choice was not only a feather in the cap of the naval engineers, by contributing one of their own, but was also a tribute to the "true and deserving worth" of Isherwood himself. Less than thirty-nine years old, he, nevertheless, had already achieved a considerable reputation for professional ability; and his appointment was a confirmation of professional respect.

Why had Secretary Welles chosen Benjamin Isherwood, junior both in age and in seniority on the list of chief engineers in the Navy? What were those personal qualities which had brought this man to the top of his profession and had won him international recognition in the field of marine engineering? How would he respond to the challenge of a position where his authority and responsibilities would expand within months to an unprecedented degree? The answers to these questions were not so simply stated.

Benjamin Franklin Isherwood was born in New York city, on October 6, 1822. His father, a graduate of Columbia College and a practicing physician in New York, died shortly after the birth of his son. Benjamin Isherwood's mother, Eliza, remarried in 1824, only to lose her second husband before her only son had reached maturity. Twice-widowed, she then remained single until her death in 1896, depending upon Benjamin for support. A strong woman, tenaciously devoted to her son, Eliza found full reciprocation in his unwavering love and concern for her welfare—a filial affection which proved to be too intense to tolerate the later competition of his marriage.

In March, 1831, when less than nine years old, Benjamin Isherwood was enrolled in the Albany Academy, a boys' preparatory school which, in many respects, was considered "a college in disguise." When Isherwood arrived at the school, the young physicist and mathematician Joseph Henry was on the faculty, although in the following year he would depart for Princeton, where he would earn increased fame for his continuing experiments in electromagnetic induction.

The transition for Joseph Henry between teaching at Albany Academy and at Princeton was not so great as it might seem. The academy, at the time Henry taught there and Isherwood attended his class, had a system of education which Henry described as being "more extensive and more thorough than that of many colleges in our country." Its curriculum, Joseph Henry later pointed out, "paralleled the courses of study at Yale College, and was more exacting in its requirements for graduation than were many of the smaller colleges." Albany Academy, in fact, had a course of study sufficiently advanced for its graduates to enter the junior and even senior years of good colleges.[3]

The emphasis upon subject matter at Albany Academy was unusual for its time. Although there was the usual attention paid to the study of the classics, the school also had shorter programs in mathematics and in the "mercantile" field. The mathematics program of study, in which Isherwood apparently enrolled, included all of the classics program, except for advanced work in Latin and Greek. In particular, his program contained courses in algebra, solid geometry, plane and spherical trigonometry, analytical geometry, and integral and differential calculus. Moreover, the student in this program, who might be barely in his teens, took courses in physics, chemistry, mineralogy, architecture, civil engineering (including topography and linear drawing), and optics.

The object of this program, established only a few years before Isherwood arrived, was to prepare the academy graduate for the practical world of business, as well as that of gracious living. The Albany Academy student, even though he might be in the "General," or classics, course, would receive above all an education of practical utility which would lay particular stress on "mechanical pursuits." Science as pure speculative theory or history as an antiquarian interest had no place at the Academy, the trustees had decided, and all courses had to serve "useful purposes of practical life," so that man and his society might benefit and progress toward perfection.[4]

Albany Academy, consequently, was a serious place with little time for frivolity. To encourage its students to prepare for the struggle of life, the academy was rigorously competitive, ranking the entire student body by class each day, awarding innumerable prizes for performance at formal public examinations, and exhorting the boys to hard work and the sober, moral life.

For the submissive or diligent pupil, Albany Academy may have been ideal. Not so, however, for those of a more positive or original bent. Into this latter category fell young Benjamin Isherwood, as perhaps also did a student in the class ahead, Herman Melville.

Under the benign despotism of the scholarly principal, Dr. T. Romeyn Beck, Isherwood and his fellow pupils labored for all but a few weeks of the year. Extracurricular activities did not exist. The boys were there to study; and under a small but well trained faculty whose scholarship often vastly exceeded its teaching skills, the students slowly advanced through the mass of material toward graduation as early as age fourteen. Benjamin Isherwood worked hard, at least initially; and in his first three years he won numerous prizes in geography, algebra, and, especially, in history.

Discipline was severe and unremitting at the academy. The principal, Dr. Beck, may have been the sensitive man whom some students recalled; but his coat of arms, as one alumnus feelingly remarked, should have been "the crimson shield, signifying gore, upon which is emblazoned the figure of a boy rampant, with the hand of one unseen holding him in position, while above, as a crest, are two rattans crossed. . . ."[5] Expulsion, according to the school regulations, might come for a wide variety of causes, in particular, "disobedience or disrespectful conduct towards teachers."[6] In January, 1836, when apparently in the final year of his course work, Benjamin Isherwood was expelled from Albany Academy for "serious misconduct."[7]

Only fourteen years old, but already possessing a formidable accumulation of knowledge in mathematics and engineering, Isherwood sought work and was hired as a draftsman in the locomotive shop of the Utica and Schenectady Railroad. For two years he remained in the shop, gaining familiarity with the structure and operations of steam boilers and engines to the point where his foreman could recommend him as well qualified to discharge the duties of a "practical Steam Engineer." After this, Isherwood spent many months in the field, absorbing the details of road and bridge construction from the British civil engineer William Lake, then resident engineer for the railroad.

From this study of railroad structures came Benjamin Isherwood's first professional publications. In March, 1842 he collaborated with another engineer to produce a pamphlet entitled, "Description and Illustration of Spaulding and Isherwood's Plan of Cast

Iron Rail and Superstructure for Railroads." Illustrated by Isherwood, this brief description of rails which had been placed in use on the Ithaca and Oswego Railroad was endorsed and highly praised by a number of railroad men, including Charles B. Stuart, an experienced engineer who later would superintend the Erie Canal and then would be engineer in chief of the Navy from 1850 to 1853.

In the following year there appeared a more ambitious work. The British engineer John Weale edited a series of articles and published the collection, in 1843, as *Ensamples of Railway Making; Which, although not of English Practice, Are Submitted with Practical Illustrations, to the Civil Engineer and the British and Irish Public*. Issued also in abbreviated form as *The Theory, Practice, and Architecture of Bridges,* this book included a long and thoroughly illustrated article by Benjamin Isherwood on the timber bridges of the Utica and Syracuse Railroad. Already bearing the unmistakable stamp of an Isherwood product, the article presented the minutest details of construction, including isometrical projections of bridges, elaborate examinations of the grading of culverts and viaducts, and exhaustive compilations of data on construction costs and timber specifications. Weale gratefully acknowledged Isherwood's work as a "liberal contribution" to the comparative study of American and British railway construction.

Styling himself a civil engineer, Benjamin Isherwood next went to work in the office of his stepfather, John Green, an engineer working on the construction of the Croton Aqueduct which would supply the water needs of New York city. At the completion of this project, Isherwood returned to the railroads, to work on the New York and Erie, under Charles B. Stuart.

Calling on the training in optics he had received at Albany Academy, Isherwood now turned for the first time to the federal government for employment, receiving an assignment by the Treasury Department to specialize in the construction of lighthouses, a duty which took him to France to superintend the manufacturing of lighthouse lenses from his own designs. Returning from Europe, Isherwood continued to work for the Lighthouse Bureau of the Treasury Department, but he soon discovered a more promising opportunity for advancement of his engineering career.

In 1842 the engineers operating the several steamships in the American Navy had finally become members of the naval service through a congressional act which permitted the Secretary of the Navy to appoint engineers to the service and to establish an engineering corps in the Navy. Perceiving the opportunities available to an experienced steam engineer in a Navy just beginning to utilize steam, Isherwood investigated the possibilities of an appointment to the corps.

In order to obtain such an appointment, however, an applicant had to demonstrate his working knowledge of marine engines. With no technical or engineering schools available to train an aspiring candidate, there was only one place to obtain the requisite knowledge and skill—the machine shops of private marine engine builders. Aided by his previous experience in railroad boiler and engine shops, Isherwood found employment with the well-known and highly regarded Novelty Iron Works, in New York city, where he gained the necessary skill with marine engines. Consequently, on May 23, 1844, he received an appointment as a first assistant engineer in the United States Navy. As this rank was only one below that of chief engineer, the highest in the corps, Isherwood was beginning his naval career with a distinct advantage.

Although appointed in May, Isherwood did not receive orders until October 1, 1844, when he was sent to the Navy Yard at Pensacola, Florida. At this pleasant, if out-of-the-way garden spot, Isherwood and his fellow officers labored under the eagle eye of their commandant, Commodore W. K. Latimer, a notorious martinet. More amenable now to discipline than in his youth, Isherwood served for a year, nursing the single ninety-eight horsepower engine and ironflue boiler of the steam tender *General Taylor*, a small side-wheeler too fragile to venture outside of harbors and landlocked bays.

Then in 1845 the first engineer in chief of the Navy, Charles Haswell, embarked on a reorganization of his corps, which had previously been haphazard in its appointments and promotions. Determined to delay no longer in rearranging his engineers on a basis of merit alone, Haswell appointed a board of chief engineers to examine all assistant engineers and to rerank them on the basis of professional and moral fitness, regardless of age, experience, or previous position in the service. Unable to go before the board in July, the time designated for examinations, Isherwood had to wait until the end of the year before traveling to Washington, D. C.

On December 18 he left Pensacola, fortified by a warm letter of recommendation from his formidable superior, Commodore Latimer. So impressed was the Commodore with Isherwood's competence that he genuinely hoped for the young engineer's return to Pensacola. Isherwood would surely receive a ranking from the board "such as your merits so justly entitle you to," Latimer believed, based on Isherwood's demonstrated efficiency, cheerfulness, and attentiveness to duties, not to mention his "strict observance of moral character and gentlemanly deportment."[8]

Unfortunately, the December climate in Washington was cold in more ways than one. After a rigorous oral testing by his superiors, Benjamin Isherwood, instead of receiving a promotion to chief engineer, found himself demoted to second assistant engineer, with his initial naval appointment revoked and his new grade warranted from January 22, 1846, thus losing both rank and seniority. For Isherwood, as well as for many of his fellow engineers, "this whole proceeding was most radical and arbitrary, and occasioned much heartburning among those unfortunates...."[9]

Isherwood had little time to brood over his misfortune. On January 26, 1846, he received orders for duty as a second assistant engineer on the steamer USS *Princeton*, which sailed from Boston in May to join the Home Squadron, which was taking part in the Mexican War by blockading the enemy coastline around Veracruz and assisting in military operations.

For Isherwood, the *Princeton* was more than just another steamer. Not only was she considerably larger and more complex than the *General Taylor*, she was one of the most remarkable warships in the world. Sponsored by Captain Richard Stockton, the Swedish inventor John Ericsson had designed this vessel which, when completed in 1843, had been the first screw-propelled steam warship in history. With her machinery placed entirely below the water line, the *Princeton* demonstrated the marked advantage of a screw-propelled warship, with no paddles or machinery to be exposed to enemy shot. By the time Isherwood was ordered to the vessel, she had already been modified by the addition of new boilers, a new propeller, and engines designed by Charles Haswell to replace the original Ericsson engines.

As in all steam warships of this period, the engines were intended to be only auxiliary to the sails upon which the ship normally depended for propulsion. Only in going in and out of port, or when becalmed, or when suddenly sent in chase of another vessel would a steam warship be expected to use her steam engines. This custom was just as well, so far as summer blockading operations on the Gulf of Mexico were concerned. During this period, Isherwood later recalled, the temperature in the engine room remained at a steady 115 degrees, while the stench from the bilge water under their feet was enough to overpower the sweltering engineers.

After spending the summer and fall of 1846 on the *Princeton,* Isherwood was ordered to a new ship, the small side-wheel steamer *Spitfire,* when she joined the squadron in November. Under the command of the peppery, impulsive Josiah Tatnall, Isherwood now found himself the senior engineer in the *Spitfire,* where life lacked even those few comforts which had been available on the larger *Princeton.*

Built originally for the Mexican government, the *Spitfire* had one small engine, which was set in crudely designed wooden frames and which relied for power on two small boilers that could only produce twelve pounds of steam pressure. The steamer was hopelessly bad under sail, normally developing as much leeway as headway in a stiff breeze. In any sort of rough weather, moreover, she was an exceedingly uncomfortable ship. With her low freeboard, the *Spitfire* readily took in the seas, which then poured into the engineers' quarters until the cabin floor was awash. Added to this discomfort, the vessel's main armament was an eight-inch pivot gun, mounted so close to the engineers' cabin that when the cannon was fired the concussion regularly shook their bunks to the cabin floor.

Throughout the winter of 1846 and into the following summer, Isherwood labored over his engine while the entire blockading force was subjected first to fierce winter storms and then to the peculiar and deadly summer pestilence of the Gulf Coast known to the sailors as the "Vomito." In March, the *Spitfire* was included in the "Mosquito division" of light steamers and gunboats which became actively engaged in bombardment operations along the coast and up the rivers into the Mexican interior. Also at this time, the *Spitfire* received a witty, courageous young first lieutenant as executive officer. His name was David Dixon Porter.

Porter, known to be "a warm friend to young officers," apparently got along quite well with Isherwood throughout the busy months while the *Spitfire* took part in such actions as the famous bombardment of the Castle of San Juan de Ulua and the river operations against Tuxpan and Tabasco. At one point, during an attack on Tlacotalpan, Isherwood left his engine to take part in a landing party which clashed briefly with Mexican soldiers. Both Tatnall and Porter complimented Isherwood on his conduct, Porter assuring the engineer, in a letter dated July 28, 1847, that "no one has exhibited more zeal than yourself in marching to meet the Mexicans."

In view of the bitterness which existed between Porter and Isherwood in the 1860's, Porter's opinion of the engineer in 1847 commands particular attention. First as executive officer and then as commanding officer of the *Spitfire*, Porter wrote warm testimonial letters on Isherwood's conduct. In one unsolicited letter of commendation, Porter complimented his engineer on the exemplary performance of the engineering crew, and went on to thank Isherwood for serving as watch officer in place of regular line officers when the latter had been unavailable. Although unqualified to judge Isherwood's professional qualifications, Porter remarked, his performance of duties had always provided such perfect satisfaction that Porter would always welcome Isherwood under his command.

Before Isherwood left the *Spitfire* in August, 1847, he received his warrant as first assistant engineer, as of July 10, thus returning him to his original appointive rank in the Navy but without his original seniority. After several months leave he was then assigned to the office of the engineer in chief, Charles Haswell; but his stay was brief, for late in February, Isherwood was sent off on duty "connected with lighthouses," which again took him abroad. While in Europe he found time to pursue his interest in steam engineering, for he sent back thorough and critical reports on proposed methods of utilizing steam more effectively as a motive force. On August 13, 1849 he finally received his promotion to chief engineer, thereby becoming for the first time a commissioned officer in the United States Navy.[10]

In late November, 1850 disagreements and maneuverings for power within his corps presented Isherwood with an opportunity for advancement and preference. The incumbent engineer in chief, in his dedication to duty, had assumed virtually the entire task of designing as well as supervising the construction of the Navy's new steam warships. After several successes, Mr. Haswell met failure with his *San Jacinto,* a strangely designed vessel even for those days. Her engines were placed far aft, so that the stern rode deep in the water, and Haswell had designed a ten-ton, six-bladed, screw propeller which was placed behind the rudder and on a shaft deliberately located some twenty inches to one side of the center line of the vessel's keel. Despite his inability, through uncontrollable circumstances, to alter this arrangement, much of which had not been his own intention, Mr. Haswell was finally replaced by another civilian engineer, Charles B. Stuart, under whom Isherwood had served on the New York and Erie Railroad. On December 3, the day after Stuart came into office, Benjamin Isherwood was detached from his lighthouse duties and ordered to Washington as assistant to the new Engineer in Chief.

Not a marine engineer, Stuart necessarily depended upon his younger associate for advice, and the orders and memoranda which issued from the office of the Engineer in Chief bore the unmistakable impression of Isherwood's views. A board of engineers was appointed to examine the Haswell machinery on the *San Jacinto,* and Isherwood soon produced the design for a new, four-bladed propeller to replace Haswell's. Lighter in weight, the Isherwood propeller was placed on a shaft which ran along the center line of the *San Jacinto,* and the propeller was moved forward to fit in front of a new rudder, also designed by Isherwood. By September, 1851, the *San Jacinto* was ready for trials.

So that he might defend his professional reputation, Haswell was made chief engineer in charge of the *San Jacinto*'s trials, but this assignment made Isherwood apprehensive. Writing hastily to his bureau chief, Charles Skinner, Isherwood urged him not to expect too much from the *San Jacinto* because the steamer still had Haswell's engines; and they were so poor that they were "a disgrace to the service and our corps," an object of ridicule, and a "standing monument of Mr. Haswell's incompetency and folly."[11] In poor health, but denied, through an administrative error, from taking necessary sick leave, Haswell attempted to comply with the orders to operate his machinery on the trial runs; but he became seriously ill, and in a fit of depression, left his ship without permission, for which act he was summarily dismissed from the service. The *San Jacinto,* after repeated failures to meet the departmental requirements, received new engines in the following year.

In the office of the Engineer in Chief, Isherwood continued to supervise machinery design and enhanced his professional reputation by contributing regularly to the scholarly *Journal of the Franklin Institute.* Most of his notes and articles were little more than compilations of machinery specifications and performance data; but he occasionally entered into controversy with such vigor that the editors of the *Journal,* in printing his replies to critics, had to omit or alter his words because of the degree of personal abuse which Isherwood had showered on those who disputed his ideas.

In April, 1851 he was detached from his duty in Washington and ordered to Sankaty Head, Massachusetts, where he was to superintend the completion of a lighthouse under construction. Finding this assignment less congenial than his busy life in the Navy Department, Isherwood sent off a letter to Stuart, in June; and a few days later was back in the office of the Engineer in Chief, this time to remain for over two years.

Once re-established in the Navy Department, Isherwood contributed some original machinery designs to the Navy, but with mixed results. First he traveled to the Gosport Navy Yard at Norfolk, Virginia, in late June, 1851, to supervise the installation of new machinery which he had designed for the steamer *Allegheny.* Replacing its underwater, horizontal Hunter paddle wheels with a screw propeller, Isherwood also rearranged the position of the *Allegheny*'s engines, so that instead of having a fore-and-aft movement, the pistons worked athwartships. Isherwood then joined the pistons to the propeller shaft by horizontal connecting rods, which extended over the shaft and then reached back from crosstails. Considered quite novel at the time, this type of engine with the back-acting motion later became a standard design for the American Navy, especially during the Civil War, and became known as the "Isherwood engine."

Unfortunately for Benjamin Isherwood, the trial run of the *Allegheny* when finally held in 1853, was pronounced "an absolute and unqualified failure,"[12] and brought forth an investigation by a board of engineers. Not only were the boilers inadequate, it was discovered, but, in particular, the engines had not been adequately braced, and the resulting vibration had broken the bed plates in the bottom of the vessel. Criticized for not providing strong enough frames to compensate for the weakness of the *Allegheny*'s hull, Isherwood tartly replied that he had been asked only to build engines, not a hull to support them. Nevertheless, he must not have dismissed such criticism, for during the Civil War his engines were attacked for their excessively heavy frames.

After sponsoring a new type of steam boiler which would not only be 40 per cent more efficient with only half the volume of the usual type, but would also be so much less expensive that its cost new would be less than the scrap value of an old one, Isherwood turned again to the drawing board and produced a plan which excited interest throughout the American engineering profession.

A small iron steamer, the *Water Witch,* had been built in 1843 for the Navy and since then had undergone continued modifications until, by 1851, the original hull had been sacrificed for naval gunnery practice while the engines were placed in a new wooden hull. This second edition of the *Water Witch* was propelled by a new feathering paddle wheel, designed by Benjamin Isherwood. Using the combined movement of an eccentric and a number of joined levers, Isherwood designed his wheel so that the paddles would enter and pass through the water while always remaining in a vertical position, the object being to avoid the inefficiency of a common, radiating side-wheel whose paddles beat the water on descent and lifted it on ascent, thus losing power. The Isherwood wheel, Charles Stuart reported, "worked admirably, without jar or noise,"[13] despite the complex structure of the paddles; and it moved the steamer through the water at a brisk eleven and three-fourths knots. Although this type of design had been used by the English and the French for several years, Isherwood's feathering paddle wheel was the first used in the United States Navy.

It was not only in the mechanical end of his profession that Isherwood labored to make his name. Close to the administration of his corps, he soon involved himself in service politics, arguing strongly for the cause of the naval engineers. In February, 1852, Isherwood, representing the chief engineers, joined Stuart and First Assistant Engineer James W. King to send a petition to Congress requesting an increase in the size of the Engineer Corps. Stressing the utter inadequacy in the present number of engineers, the three insisted on the "strong probability" that in twenty years there would be no naval vessels "unpropelled, in whole or in part, by steam." Necessity would compel the use of steam power for all marine war purposes; and to meet this inevitable growth of the steam Navy, they argued, there should be a commensurate growth in the corps of naval engineers.[14] From this point, Isherwood's role as spokesman was to flourish, just as the corps itself was to mushroom in size within a few years.

Like all young naval officers, Isherwood was supposed to have occasional tours of sea duty. In March, 1854, he was ordered to the USS *Massachusetts,* which departed in early July for a lengthy cruise in the Mediterranean. Within two weeks, Isherwood was no longer with his vessel, having been removed in "critical condition" shortly after the ship had put to sea. He had come down with a case of dysentery so severe that the medical officers and the captain hastily agreed that their Chief Engineer was "very *dangerously* ill," and would not last in the "low latitudes." Wasting no time, the *Massachusetts* put in at Fayal, in the Azores, where Isherwood might recover sufficiently to be shipped back to the United States.[15]

Once home, Isherwood took months to shake off his illness; but by the following April, he was ready for his next assignment at sea. In September he received his orders; and reporting to Mr. Haswell's "folly," the *San Jacinto,* Isherwood embarked on a three-year cruise to the Far East which would be the last sea duty of his naval career.

The *San Jacinto,* with Commander Henry H. Bell as her captain, was to be the flagship for Commodore James Armstrong, the commander in chief of the East India Squadron. Rated as a second-class screw steamer, the *San Jacinto* carried a complement of eight engineers, under the charge of Isherwood, the only chief engineer aboard the vessel. The cruise, as it turned out, was to be far from routine, since, as flagship, the *San Jacinto* was to sail to the Strait of Malacca to pick up the American diplomat Townsend Harris to transport him first to Siam to negotiate a new trade treaty, and then to Japan, where he would become the first American consul general. Consequently, the ship carried a remarkable cargo of gifts to illustrate the manufacturing skill and ingenuity of the American people to the court of Siam.

The long, leisurely trip across the Atlantic to the Indian Ocean was particularly boring for the engineers and their fireroom crew because the Captain, unwilling to expend the coal, rarely used the

## The New Engineer in Chief    17

engines. Boredom grew into active discomfort as the weeks went by; for Commodore Armstrong, in an apparent attempt to inspire greater obedience and respect on board, senselessly rationed water to the point where the crew refused to eat their rations, since cooking required too much water.

To the relief of many, the *San Jacinto* finally arrived at Simon's Bay, South Africa, on January 12, 1856. While the ship remained in port for two weeks, Isherwood wasted no time in visiting all of the British steamships in the harbor to examine their machinery and to compile records of their performances for his ever growing files.

On March 21, the *San Jacinto* arrived at Pulau Penang, in the Strait of Malacca, where Townsend Harris came on board for the trip to Bangkok, where they arrived the middle of April. Greeted there by a small, sky-blue steamer, ambitiously called *The Siamese Steam Fleet,* the officers of the American warship accompanied Townsend Harris up the Me Nam River on his visit to the King of Siam. Surrounded by the trappings of oriental splendor and borne on sedan chairs past a remarkably heterogeneous palace guard which included twenty elephants, each with an ancient howitzer on its back, the dazzled Americans came before the august presence of the ruler before whom the Siamese nobility had already prostrated themselves.

Surfeited by royal hospitality and successful in their mission, the Americans left Siam on May 31, and sailed toward the East India Squadron headquarters at Hong Kong. After some engine trouble, the *San Jacinto* arrived there on June 12, to be greeted by a swarm of Chinese, including one enterprising and knowledgeable boatman who proudly sailed up to the American warship, flying from his masthead a large flag which read, "BUM BOAT. U.S. STEAMER SAN JACINTO."[16]

Engine trouble continued to plague the ship, so that it was not until July 10 when the *San Jacinto* finally sailed for Japan to deliver Townsend Harris. Scarcely had the steamer traveled one mile, however, before the keys fastening the propeller to the shaft broke, and the whirling blades slipped back into the rudder, nearly taking the sternpost off the vessel. Towed to Whampoa, "a miserable, marsh-surrounded, pestilential anchorage,"[17] the *San Jacinto* was in dry dock for two weeks before her engineers could finish their repairs. On August 10 the steamer once more set out for Japan, this time arriving, on August 22, at Shimoda, where Townsend Harris left the ship.

The *San Jacinto* returned to the naval depot at Shanghai in September, to find sporadic warfare going on between the Chinese and the English. In November, the crew participated in the bombardment of Chinese positions at Whampoa. Isherwood, however, was quite unable to participate; for, in July, he had, once again, come down with a severe case of chronic dysentery.

Exhibiting the classic symptoms of this disease—severe cramps, nausea, and a tongue "quite covered with a long, dirty-white fur"— Isherwood submitted reluctantly to the ministrations of the *San Jacinto*'s assistant surgeon, R. P. Daniel, who stuffed the engineer with stiff doses of lead and opium, supplemented with opium suppositories and a diet of rice gruel. Never one to obey a doctor's admonitions, Isherwood would consent to this regimen only when seriously ill; once improved, he would immediately go on a gastronomic spree and then suffer the inevitable relapse, while the doctor noted in exasperation that it "appears impossible to curb his appetite within the bounds of a proper diet, either in quantity or quality."[18]

Failing to recover his health after a visit to the relatively bracing climate at Macao, Isherwood was put ashore at Hong Kong to rest from the middle of November, 1856 to the following January. He rejoined his ship just in time to avoid being poisoned by a group of xenophobic Chinese who had determined to erase the foreign population by mixing arsenic in the fresh bread. The *San Jacinto* fortunately patronized a bakery other than the one "through which the mischief was done," but hundreds of Europeans were not so lucky, and although only one person died, the digestive apparatus of many was never again quite the same.[19]

To the utter mystification of the ship's doctor, Isherwood still catered to his "huge appetite, which he appears to gratify with impunity," and by the spring of 1857 he was sufficiently recovered to return to duty. Despite occasional relapses, he continued "in good flesh," and once the *San Jacinto* finally left the miasma of the Chinese mainland, he was free from the disease.

That Isherwood had been rash in the cavalier treatment accorded his stomach was all too evident to Dr. Daniel. Dysentery in the 1850's was a formidable disease in the American Navy. Unless controlled within the first five or six weeks, the doctor observed, "it almost invariably terminated either in death, or in a chronic condition which baffled our every effort to produce a permanent cure"—and Isherwood's case took two years to bring under control.[20]

Returning to the United States in August, 1858, Isherwood entered a new phase of his naval career. Throughout his cruise in the *San Jacinto* he had filled his leisure time, and had avoided the lethargy induced by climate and food by a ceaseless pursuit of scientific knowledge, which he equated with the remorseless collection of every available scrap of engineering data. Aware of his preoccupation, his superiors placed him on many experimental boards during 1859 and 1860 so that the Navy might utilize the encyclopedic knowledge that Isherwood had at his command.

In September and October, 1860, he took part in a department survey of all the sailing vessels in the Navy which intended to determine how many might be converted to steamers. Isherwood and his associates concluded that the smaller warships—brigs, sloops, and frigates—should retain their full sail power; but ships of the line were now useless and should be razeed and converted into first-class, screw-steam frigates.

In November, 1860, Isherwood received an assignment which proved to have the greatest significance for his career both as an engineer and as a naval officer. At Erie, Pennsylvania, the old paddle-wheel steamer *Michigan* became destined for engineering fame—or notoriety—as a board of engineers under Isherwood's direction spent the winter months experimenting with her engines. Their assignment was to ascertain the most economical method of using steam in the reciprocating steam engine of that period. They were aware that the savings in fuel costs in steam generation by allowing expanding steam to do much of the work within a cylinder were theoretically great. Isherwood's board, including Chief Engineers Theodore Zeller, Robert Long, and Alban Stimers, labored to determine just how economical the use of expanding steam really would be, by balancing the savings in the cost of coal burned in generating the lesser amount of steam against the loss of steam pressure through expansion and the consequent loss of engine power.

The report of the board on February 18, 1861, by challenging the normal practice as an excessive use of expanding steam, fanned an issue already smoldering in engineering circles. By his prior work in this area and his championing the controversial conclusions in the Erie report, Benjamin Isherwood boldly thrust himself and his theories of steam engineering in the face of doctrines accepted by the great majority of steam engineers, both here and abroad. As a result of his experiments, he became the best known and most controversial engineer in the Navy, if not in the United States.

It was also at Erie where Isherwood made a friend of Theodore Zeller who would be his devoted associate and intimate companion for the next forty years. Polished in his manners, scholarly and refined in his address, and quietly exuding the breeding of generations of New York city aristocracy, Theodore Zeller made an interesting contrast to the brusque, energetic Isherwood. The two shared a love of music and art and a devotion to their profession. Here the similarities ceased. Zeller was ever agreeable, with a placid disposition which covered his personality as a blanket. With his "singularly even" temper and charitable instincts, which undoubtedly proved at times to be his undoing, Zeller amiably accompanied his dynamic and impatient friend through the years of naval and engineering life. Isherwood always led; Zeller never failed as the loyal follower, whose unquestioned acceptance of his friend's superior energy and ability kept the two bound by a symbiotic relationship severed only by Zeller's death.

Detached from his work on the *Michigan*, Isherwood went to Washington in the middle of February, 1861, to be on a board investigating the cause of oxidation in the boilers of the USS *Dacotah*. Within the month, Engineer in Chief Samuel Archbold suddenly resigned both his position and his commission as a naval engineer and left the service, for reasons which were apparently personal rather than concerned with the political unrest which already threatened the solidarity of the Navy.

If Isherwood solicited the position vacated by Archbold, there is no evidence of such an effort. Undeniably, he had friends in the right places, especially John Lenthall, the chief of the Bureau of Construction, Equipment, and Repairs. Lenthall, a civilian and considerably older than Isherwood, had worked closely with the young engineer while on duty in Washington, and the congenial relationship between the two was to ripen over the years into a close friendship—one of the very few Isherwood permitted himself.

It would be logical for Gideon Welles, secretary of the Navy for only a few days, to have depended heavily on Lenthall's judgment. In any case, at this crucial period the Navy needed a man of proven energy and ability who would be able to handle the staggering amount of work which might suddenly be thrust upon him if the Union were to commence naval operations against the South. Gideon Welles found that man in Benjamin Isherwood.

Within only a few weeks, Secretary Welles had ample opportunity to judge the merits of his new Engineer in Chief, and then it was at a moment of national crisis. As Virginia teetered on the brink of secession, the Gosport Navy Yard at Norfolk assumed a critical role in political and military affairs. Unwilling to antagonize the residents of the surrounding countryside and precipitate secession by forcing overt southern resistance, President Lincoln and his Secretary of the Navy grudgingly maintained the *status quo* in the Norfolk yard, neither building up its defenses nor removing any of the significant portion of the American Navy which was anchored within the yard.

In command of the Norfolk yard was the venerable Commodore Charles E. McCauley, who had earned this comfortable billet by serving fifty-two years in the Navy. With a career stretching back to the days of Isaac Hull and Stephen Decatur, McCauley had long outlived his usefulness as an active seagoing officer. Now, in his declining years, little more than a symbol of naval tradition, he found himself suddenly enveloped in an "atmosphere of treason," surrounded by younger officers who were largely secessionists and shaken by a series of anonymous threats which made violence appear imminent. McCauley, far more than Gideon Welles, felt the need for caution—even to the point of inaction.

By early April, Secretary Welles believed he could wait no longer. Although building land defenses against the citizens of Norfolk would be open provocation, the careful removal of a warship might not unduly disturb the existing situation. Consequently, he decided to save the *Merrimack,* a vessel far more valuable than any other in the yard and a vital part of the small Union Navy.

The *Merrimack,* a forty-gun, screw-steam frigate, had been designed by John Lenthall and completed in 1854. A wooden-hulled vessel, primarily intended for cruising under sail, her relatively small and inefficient engines were at the moment dismantled and were being repaired under the supervision of Chief Engineer Robert Danby, at the Norfolk yard. Through normal departmental correspondence, Benjamin Isherwood was aware of the steamer's condition, but he was also aware of the situation in and around the yard. Realizing the danger of the *Merrimack* in southern hands, Isherwood went to see Welles about saving the ship.[21]

The extent of Isherwood's influence on Welles at this point cannot be determined, but the Engineer in Chief was undeniably insistent in stressing the importance of the *Merrimack,* and he "repeatedly urged" the Secretary to rescue the steamer. By April 10, Welles decided to remove the *Merrimack* to the Philadelphia Navy Yard, and ordered her preparation for this trip "with the utmost despatch." The reply from Norfolk threw Welles into consternation. The engine repairs on the *Merrimack* would take at least four weeks; the vessel was helpless until the repairs were done. Isherwood, however, flatly rejected this estimate as being far too pessimistic. The work, he believed, could be done in a week; the four-week estimate was an excuse to keep the *Merrimack* in reach of southern sympathizers. The only recourse, Welles realized, was to send Isherwood himself to Norfolk to supervise the repairs.[22]

On April 12, Isherwood set out for Norfolk, together with Commander James Alden, who would sail the *Merrimack* to Philadelphia once she was ready for sea. Arriving in Portsmouth, Virginia, on the fourteenth, Isherwood located Chief Engineer Danby, and the two went to Commodore McCauley's office to present the old man with orders from Welles requesting his co-operation with Isherwood's work. McCauley readily consented, and the two engineers then boarded the *Merrimack* to examine the task before them. "The engines," Isherwood recalled, "were in a wretched state." The engine braces had been removed, and the machinery was scattered throughout the yard. However, disabled as the *Merrimack* was, there was still hope, since the machinery was all there. What was needed now was an extraordinary amount and speed of reassembly.[23]

First recruiting a number of machinists to replace the local laborers who had quit in an attempt to delay the work, Isherwood divided his labor force into three eight-hour gangs to work around the clock, and then steadily urged them on, day and night, with no respite. The driving, uncompromising spirit of the Engineer in Chief brought forth a herculean response from his men. Whether goaded by his impatience or inspired by his fervor, they worked ceaselessly until the job was done. On Wednesday afternoon, two and one-half days after the work began, they had repaired the *Merrimack*.

Exhilarated by their accomplishment, Isherwood and Danby reported to McCauley that the *Merrimack* had received fuel, stores, and a special crew of forty-four firemen and coal heavers personally hired by Isherwood; as soon as McCauley gave permission to fire up, the steamer would be ready to steam out of the port. McCauley, however, saw no need for immediate action; tomorrow morning, he believed, would be early enough for raising steam. Isherwood returned to the ship and put on the regular engine-room watch, ordering them to light fires shortly after midnight.

Promptly at nine o'clock Thursday morning, Isherwood returned to McCauley's office to inform the Commodore that the crew was on board, the steam was up, the engines were working, and all that remained was McCauley's order to cast loose the *Merrimack* and take her to safety. To Isherwood's "great surprise and dissatisfaction," McCauley stated that he had not yet decided about sending the *Merrimack* to Philadelphia at all; he would let Isherwood know in a few hours. Astounded at this apparent disregard of both Welles's orders and the obvious danger to the *Merrimack,* Isherwood reminded McCauley of the peremptory nature of Welles's instructions. Futhermore, he said, the *Merrimack* could now pass over any obstructions in the channel, but any delay might permit the southerners time to block the exit. McCauley stolidly replied that he would make up his mind at a later time.

Throughout this conversation and in subsequent ones, Isherwood later recalled, McCauley had appeared to be completely prostrate, immobile, stunned by the nature of the crisis, and apparently befuddled from drink. Overwhelmed by his unaccustomed responsibilities, wavering at a time for decision, the aged Commodore was tragically incapable of effective command.

Realizing the futility of moving McCauley to action, Isherwood turned to Commander Alden. The engineers' work was done, Isherwood explained, showing Alden that the *Merrimack* was ready; now it was the duty of the line officer to take command of the vessel. Once again, however, there was a lack of that leadership and initiative so necessary in a crisis. Alden, at first enthusiastic about removing the *Merrimack,* now lost his ardor and shrank from the responsibility of overriding the Commodore's authority and removing the *Merrimack* on his own.[24]

Returning to McCauley, Isherwood was further dismayed to discover that the Commodore had finally made up his mind; the *Merrimack,* McCauley now insisted, could not escape, for the channel had already been blocked. There was nothing else to do but stop the engines, extinguish the boiler fires, and keep the *Merrimack* at Norfolk. Since the Commodore would not listen further to Isherwood's arguments that the warship could still be moved with perfect safety, there was nothing left for the Engineer in Chief but to return to Washington and inform Welles of the tragic situation.

Angered by McCauley's moral paralysis and Alden's dread of assuming authority, Isherwood considered rescuing the vessel himself. He had kept the engines running continuously to demonstrate that the vessel was ready to leave. He had personally hired a crew to operate the ship. Without authority he had removed the chain cables binding the *Merrimack* to the dock, and had substituted rope hawsers—and placed men with axes to cut the ropes when he gave the signal. With few stores and no armament, the ship would ride high enough in the water to pass over any obstructions now in the channel. When McCauley ordered him to stop the engines and draw the fires, Isherwood recalled, he was "greatly tempted to cut the ropes that held her, and to bring her out on my own responsibility."[25]

However, such a seizure of the ship without authority and against orders would have been an unforgivable act. The laws of the American Navy were very clear on this point; no engineer could encroach on the prime prerogative of line officers, that of command. Rescuing the *Merrimack* would have demonstrated great initiative and daring, but it would also have been a most serious breach of naval discipline. Isherwood also realized, looking back over the affair, that an adequate excuse for such unauthorized action did not exist at the time, but only emerged later with the dramatic consequences of the Confederate seizure and conversion of this steam frigate into the ironclad *Virginia.* Yet had he rescued the *Merrimack,* "the disasters which followed her detention, and which are my justification for the desire to take the matter into my own hands, would not have happened,"[26] thus forcing him to rely only on

*The New Engineer in Chief* 25

the saving of a fine warship to explain his actions—an insufficient excuse for the Navy.

Isherwood prepared to return to Washington, leaving behind the pride of the Navy's steam fleet, although by this time Virginia had seceded from the Union and there was no further need to placate her citizens. Still unable to act, Commodore McCauley remained at his post, dimly hoping for the reinforcements which would never come.

Meanwhile, secessionist elements in Norfolk, outraged at Isherwood's attempts to snatch the coveted *Merrimack* from their grasp, determined to capture the Engineer in Chief and to hold him as a prisoner of war. Fortunately for Isherwood, a sympathetic friend in Norfolk discovered the plot, warned the Engineer, and arranged for his secret departure. Obtaining a cabin in his own name on the regular Chesapeake Bay steamer, Isherwood's friend boarded the ship with the Engineer in Chief's trunk, then slipped ashore and rode to the Atlantic Hotel where Isherwood was smuggled into his carriage, which then returned to the ship. Boarding inconspicuously, Isherwood locked himself in the cabin until the steamer was safely on its way to Washington. The party of Confederates assembled on the wharf to capture him waited in vain and discovered the deception only after they had returned to the hotel in search of Isherwood.[27]

On the morning of April 19, the Old Bay Line steamer arrived in Washington where Isherwood and Alden, who had also decided to leave Norfolk, reported to Welles. Furious at Alden's failure to rescue the *Merrimack,* the Secretary still hoped to discover a way by which the Navy Yard might be defended, but subsequent discussions with military advisors brought Welles and the President to the reluctant conclusion that the yard must fall to the Confederates.

At this point, McCauley, finally despairing of any help and expecting his yard at any moment to be overrun by Virginian troops, took the situation into his own hands. He promptly scuttled all the vessels in port, including the *Merrimack,* whose engines and boilers, however, were still intact. Shortly afterward, a Union naval force led by Commodore Hiram Paulding arrived at the yard and attempted to destroy the entire installation; but when the Confederate forces moved in they were able to seize not only the smoldering and partly submerged *Merrimack,* but also a large amount of cannon and naval stores which would later prove to be of immeasurable value to their cause.

Responsibility for the loss of the Navy Yard rested on many shoulders. Welles and Lincoln had delayed initially through fear of provoking secession. McCauley and Alden, the senile Commodore and the Commander with "heroic drawing room resolution and good intentions," had vacillated and retreated into impotent inaction, although one might argue that they should have received more explicit and forceful orders from the Navy Department. Only Benjamin Isherwood emerged from the episode with his reputation entirely untarnished. In his circumscribed field of action, he had moved with notable speed and decision. Furthermore, he made a vigorous effort to the limits of his authority to rescue the *Merrimack.* Failing in this endeavor, he had no alternative but to return to Washington.

Isherwood's importance in this episode rests on several historical might-have-been's; for the successful removal of the *Merrimack* would certainly have altered the course of naval operations in the Civil War, and might have drastically changed the course of American naval technology and strategy for a generation. Without the challenge of the Confederate ironclad built on the hull and machinery of the old *Merrimack,* John Ericsson's revolutionary ironclad *Monitor* might not have appeared or engaged in the dramatic encounter at Hampton Roads. This battle, so influential in encouraging the development of ram tactics and in starting the American Navy off on a "monitor craze" which proved to be a strategic handicap in later years, might never have occurred without the effective challenge by the formidable Confederate ironclad which, phoenix-like, had risen from the ashes of the *Merrimack.*

Ironically, Isherwood was to a degree fortunate in his failure to prevent the loss of the *Merrimack.* His actions in this situation demonstrated that energy and resourcefulness which Welles had sought in choosing his new Engineer in Chief. Free from any blame for the *Merrimack*'s loss, Isherwood had earned the respect and confidence of his superior; and by wisely acknowledging the limitations of his authority, he had avoided a bitter conflict with officers of the line at a time when such conflict would have shattered the effectiveness of the Union Navy.

## II. Building the Union Navy

"The first act I had to perform as Engineer-in-Chief," Isherwood later recalled, "was to prepare machinery for the war service, as the war was then upon us."[1]

At the beginning of March, 1861, the American Navy consisted of ninety vessels, of which forty-two were in commission. There were only twelve stationed in the home squadron, and of these, just four ships, with twenty-five guns among them, were in Northern ports.[2] To hastily reassemble a fleet spread all over the world was a difficult task for the Navy Department; far more challenging would be the effective use of these few ships.

The most obvious role for the Navy at the beginning of the Civil War was to blockade southern supply routes. This was a staggering undertaking. From Alexandria, Virginia to Brownsville, Texas stretched 3,550 miles of coast, much of which was "double coastline," providing a continuous, shallow inner waterway in which Confederate blockade-runners and coasters could operate with relative impunity. By his institution of a formal blockade of the South on April 19, 1861, President Lincoln gave his Secretary of the Navy the task of patrolling 189 harbors and navigable river mouths with a handful of naval vessels. If it were successful, this "Anaconda Policy" of strangling the South through disrupting its trade and economy would devastate the Confederacy by attrition, while complementing the military strategy of the Union Army.[3]

To achieve a successful blockade, Secretary Welles had to improvise a large fleet of steamers, capable of staying continually on station without breakdown, and fast enough to discourage all but the swiftest Confederate blockade-runners. For such warships, Welles once again turned to his Engineer in Chief.

Benjamin Isherwood, in 1861, presented an appearance and manner which alone were convincing evidence of his ability to meet the Secretary's most stringent demands. Isherwood was a heavy-set man, with immensely broad shoulders and a thick chest. Although five feet, ten inches in height, he did not appear to be that tall because of his massive frame; yet his quick movements and alert manner dispelled any impression of corpulence. And the combination of his dark, masculine features, curly hair, and sensitive, if often forbidding countenance, won him an accolade as "the handsomest man in Washington" during the Civil War.[4]

Although reputed to be a great conversationalist, especially with the ladies of the Capital, Isherwood was not the charming *bon vivant* his well-meaning friends have portrayed. Passionately devoted to thoroughness and accuracy, and indifferent to his personal reputation with others, Isherwood was not an easy companion. His enormous energies far exceeded those of most of his associates; and his basic impatience, under the pressure of the war years, grew into an unremitting intolerance of stupidity, laziness, and error. He had neither time nor desire to cultivate people, regardless of their importance, and this indifference too often appeared to be contempt. He was admired, but all too often grudgingly, and his consistently unequivocal position on any issue, controversial or not, won him few real friends, but many dedicated enemies.

To create a steam Navy, Isherwood could draw on few resources other than his own abilities, but he received invaluable aid from a colleague. As Engineer in Chief, he was still under the immediate supervision of the Bureau of Construction, Equipment, and Repairs, currently directed by the civilian naval constructor John Lenthall. When he became head of the newly created Bureau of Steam Engineering, in 1862, Isherwood continued his close association with his former Chief. The Union and its Navy were fortunate that Lenthall and Isherwood were fast friends and would remain so throughout the war. Lenthall, fifteen years older than Isherwood, had a disposition closely resembling that of his colleague. Without, as Gideon Welles recorded in his diary, "much pliability or affability," John Lenthall was sternly honest, and as dedicated to his profession as was his young engineering associate. His "unaffected manner ha[d] offended others" just as Isherwood's tactlessness had endeared him to few, and Lenthall's indifference to private interests won him a bitter, destructive criticism in which Isherwood was soon to share.[5] With unfaltering faith in each other's abilities, Lenthall and Isherwood worked together in complete trust, meeting the continuing challenge of technological and naval developments with confidence and conviction.

Totally out of keeping with Benjamin Isherwood's personality were the surroundings in which he labored. In 1861 the Navy Department was housed in a "small red-brick building ... very plain and even humble." In this building Isherwood's offices were located in "rooms below stairs," while the Secretary and his staff enjoyed offices which lined both sides of an upstairs hall. To lend a feeling of tradition, a number of oil paintings, water colors, and engravings adorned the walls along the upstairs corridor.[6]

Isherwood started with a small staff which expanded only slightly throughout the war. This office, initially part of Lenthall's bureau, contained several young naval engineers assigned to assist the Engineer in Chief. As the increasing amount of detailed and largely mechanical work threatened to inundate the engineers, Lenthall, in June, 1861, requested both a clerk and a draftsman for Isherwood's office, noting that there were none attached there at the time.

Among the young engineers who worked with Isherwood, no one questioned the Engineer in Chief's absolute authority. According to Clark Fisher, a naval engineer who worked with him during the Civil War, Isherwood "instilled a wholesome dread of any carelessness or error" in his subordinates. Not only was he "quick to discover a mistake," but he was "never at a loss for incisive words expressive of his opinion of the culprit and his work, or lack of it."[7] His dedication to exact, exhaustive work easily explains the impatience and abruptness which kept his juniors full of a healthy respect for their superior.

Yet Isherwood was not a martinet merely seeking to gratify his cravings for power by a ruthless rule. If he was "unsparing in criticism," he was also "generous in approval and commendation." His professional dedication to both engineering and naval progress had a contagious quality, capable of driving his subordinates to a degree of effort and enthusiasm they had never anticipated. The virtues of and necessity for teamwork were not lost to Isherwood, for "his assistants were always made to feel rather as associates, engaged in advancing a common cause than as subordinates, receiving and executing orders from their superior officer." In this manner Isherwood drew "zealous co-operation" rather than "perfunctory service" from his men, and the office of the Engineer in Chief operated with an efficiency and a productivity which contributed heavily to the success of the steam Navy throughout the Civil War.[8]

The magnitude as well as the difficulty of Isherwood's task was unprecedented. With only a handful of steamers, the American Navy suddenly had been called on to perform a function which required hundreds of vessels, most of which had to run by steam power. To cope with this problem, Isherwood had no traditions and no precedents in creating such a steam fleet. America had never been a big naval power; and, traditionally, had relied on a last-minute collection of whatever ships were available to meet wartime needs. In the Civil War the Navy was as unprepared as ever; but, in addition, it now had the problem of producing warships which had to utilize a relatively new and unpredictable form of propulsion. Never had a naval engineer been faced with this situation, for never before had large-scale naval operations had to rely on steam engines. Throughout the Civil War, all that Isherwood did to design, build, and maintain the motive power of the steam fleet would be unprecedented. It was inevitable that he would make mistakes; the department only hoped that he could meet the challenge.

The most pressing need of the Navy, when Isherwood became Engineer in Chief, was for small gunboats which could be used for close, inshore work and for supporting amphibious operations against Confederate strongholds, such as Port Royal, South Carolina. In March, 1861, there were none of these vessels in the Navy, but Isherwood quickly supplied an answer to the problem. He had recently been employed by the Russian government to design the machinery for two 691-ton gunboats to be used on the Amur River, under the command of the Russian Captain Davidoff. Isherwood not only designed the engines, but also superintended construction of these boats at the Novelty Iron Works, in New York city. When Welles requested the immediate construction of gunboats, Isherwood was able to present him with complete plans of the "Davidoff Gunboat."

He recommended that Welles immediately contract with the Novelty Iron Works for four such vessels, since the drawings, specifications, and patterns were complete. In addition, the cost of such vessels was known exactly, since Davidoff had visited nearly all the principal American machine shops looking for the lowest bid. Welles showed little enthusiasm for this suggestion, wishing instead to follow the normal government contract procedure of advertising for the gunboats. Isherwood and Lenthall, however, unwilling to waste such a unique advantage, finally persuaded Welles to contract directly with the Novelty Iron Works.

As a result, the first four 9½-knot steamers were built in the phenomenally short time of 90 days, thus earning this class of vessel the name of "ninety-day gunboats." Ultimately, 23 of these small, heavily armed, screw vessels were built, all of which had the same Isherwood engines, although the last 19 had 60 per cent greater boiler power.[9] With nearly all the contracts for these gunboats let in early July, 17 of them were in active service by the end of the year, proving to be of great value to the Union Navy in its early operations.

The next class of vessels built under Isherwood's supervision was an unusual type, called forth by the exigencies of river operations against Confederate forces. Inland waterways so sinuous that turning a vessel around would be hazardous if not impossible required speedy, shallow-draft boats which could go with equal facility in either direction. For this purpose the Navy decided on a class of side-wheel "double-enders," of which 12 of the 1,100-ton, 11-knot vessels with rudders at both ends were built, starting in the summer and fall of 1861. These gunboats all used an orthodox, direct-acting inclined engine of Isherwood's design, and also employed the new feature of forced draft by the use of mechanical blowers.[10] In the autumn of 1862 the construction of an improved class of double-enders was under way, these *Sassacus*-class vessels having a designed speed of 14½ knots, with 4 more double-enders, featuring iron hulls, being able to attain 15 knots.[11] These gunboats, displacing 1,173 tons, were considerably larger than the earlier class of double-enders which were about to join the fleet.

Not only were small gunboats needed in great quantity. For cruising at sea, the Navy also required fast screw sloops, able to intercept the coastal blockade-runners which infested southern waters. Fortunately, Congress had authorized the construction of 7 sloops of war in February, 1861. The department, using the plans of sloops built in 1858, constructed 4 of these 1,560-ton vessels, including the *Kearsarge,* destined for fame in its fateful encounter with the Confederate raider *Alabama.* Congress also authorized the construction of 10 additional sloops of war at this time, and these were equipped with Isherwood's machinery.[12] Larger than the 1858 models, the new screw steamers varied in size, ranging from 1,934 tons to 2,200 tons, and producing speeds between 12 and 13 knots in the open sea. Construction began early in the fall of 1861.

Near the end of 1862, as Isherwood was working busily upon his improved class of double-enders, additional contracts were let on a class of 12½-knot, small screw sloops. This group of 8 vessels, which included the *Nipsic,* was intended for cruising close to the shore, as the sloops were only 150 tons larger than the ninety-day gunboats, and were built mainly for speed.[13]

In addition to the close-in blockade of the southern coastline, the Navy also had to man the "outer line," approximately one hundred miles offshore, where large, swift cruisers were stationed to cut off the larger blockade-runners and to challenge the formidable Confederate commerce destroyers which roamed the open seas. For this purpose, the Navy Department, in 1863, projected a group of twenty vessels, including both gun-deck frigates and sloops of war. All of these ships were to be built with two-cylinder, back-acting engines of Isherwood's design. Eight of them, the large frigates, were to displace approximately four thousand tons and have a thirteen-knot speed. Best known of this class were the *Antietam* and *Guerriere,* both serving in the Navy for many years. The smaller sloops of war, of the *Contoocook* class, displaced only three thousand tons, and, consequently, were speedier, being designed for fifteen knots. Of the ten proposed, only four of this class were built.

Finally, in 1863, the Navy Department, with an eye toward possible British and French intervention on the side of the Confederacy, planned a class of supercruisers in which all other qualities were to be subordinated to speed. These vessels, wooden hulled, unarmored, and carrying enough sail power to permit extended periods of cruising without steam, were the result of lessons imperfectly learned from the War of 1812. The aura surrounding the exploits of America's superfrigates, such as the *Constitution,* led American naval strategists of the 1860's to believe that if war with a large commercial power once again came, America would repeat the terrorizing raids of its legendary frigates, and that such exploits would bring the enemy to its knees. Thus there came into being a class of vessels which never took part in the Civil War, but which nevertheless vitally affected the naval and engineering reputation of Benjamin Isherwood. The brief careers of the *Wampanoag* and her sister ships sounded, in the later 1860's, a tragic note which would produce echoes in subsequent years as the Navy made its descent into the "dark ages" of American naval history.

Between 1861 and 1865, Benjamin Isherwood designed the machinery for forty-six paddle-wheel vessels and for seventy-nine screw steamers. Not only did he produce the general plans for such machinery, but he spent sixteen to twenty hours a day in his office turning out "the most minutely detailed specifications" and all the working drawings for constructors.[14] The ingenious inventors of his day could also turn out engine designs quickly enough, but Isherwood had a further responsibility. His engines had to run, and they had to continue running without breaking down.

The tremendous increase of the Navy's steam fleet during the war posed many problems, not the least of which was finding personnel capable of handling the machinery in steam vessels. As patriotic young men flocked to the Navy Department to offer their engineering talents, zeal soon outdistanced skill as the major contribution of engineering recruits. As much as the Engineer in Chief labored to screen the applicants and train these new men, it was not enough. There was no naval engineering school to teach them their trade; they had to learn by experience, and often while under fire. The best way to deal with this problem was not to spend precious weeks and months training the engineers, for there was not time for that. Instead, as Isherwood soon realized, he would have to design and build his engines to be so simple and reliable that any novice could operate them without causing an immediate engine breakdown.

With durability and reliability as his guiding principles, Isherwood designed engines which could withstand the manhandling of the clumsiest of mechanics. His engines were immensely strong, and, consequently, immensely heavy. Fuel economy and power, though desirable qualities, took second place to simple mechanical dependability. In a period of rudimentary steam technology, when low steam pressures were necessary, Isherwood refused to employ more advanced theories which utilized high pressures and great degrees of steam expansion. The success of engines built on more sophisticated principles was too questionable, considering the low caliber of most engineers and the tremendous demands placed continuously on naval steam engines. His engine designs, therefore, became an easy target for those who criticized Navy steamers for lack of power or economy and for the excessive weight of their machinery.

Isherwood's requirements for engines clearly demonstrated his absorption with practicality in machinery design. Insisting on "fairness of parts" in his engines, he stressed simplicity of combination and an arrangement which would feature easy access and constant observation. All surfaces in moving contact with other metal parts were to be large in order to withstand the great strains and continuous abrasion in a working engine. Anticipating the inevitable mechanical problems caused by bad materials and bad workmanship in construction, as well as from mismanagement and abnormal strains through continuous operation, he insisted on extraordinary strength of components.[15]

At times, Isherwood's best efforts to forestall engine breakdowns failed. Finding competent engineers when so many were needed was not always possible, as he soon discovered. In July, 1863, Admiral Farragut, in the midst of directing naval operations at the mouth of the Mississippi River, wrote testily to Welles about the poor condition of the new vessels being sent out to him, ascribing the difficulties to the rapid increase in the Engineer Corps. "The majority of them know very little of their duties," the Admiral said, "and their engines are cut up and ruined by neglect and want of proper care."[16] Welles turned the problem promptly over to Isherwood, remarking that the difficulties apparently stemmed from "employment of incompetent and neglectful persons in the engineer department." He urged Isherwood to attempt to improve the caliber of the Engineer Corps, especially of the temporary branch comprised of noncareer marine engineers being utilized only for the duration of the war. To safeguard against the employment and appointment of incompetents, Welles informed Isherwood that the department would send a circular letter to all Navy Yard commandants, encouraging a more rigid screening of applicants.[17] Isherwood, perhaps more realistic in his view of the situation, continued to design more rugged and fool-proof engines.

Bad workmanship in engine construction was a problem scarcely less serious than that of mishandling by inexperienced engineers. Private contractors were unprepared for the deluge of orders suddenly thrust upon them during the war, and inevitably the quality of their product deteriorated as they struggled to complete their work against rising material and wage costs which threatened to wipe out the handsome profits they had at first envisaged. The likelihood of hidden defects in the hastily made forgings produced by such contractors forced Isherwood, in designing his engines, to use an enormous factor of safety to forestall wholesale engine failures caused by breaking parts.[18] This practice, along with the need to protect the engines from the engineers, forced Isherwood into a vulnerable position, and he did not have to wait long for the onslaught of criticism. Yet his engines worked and worked reliably, and that was all that really mattered. Despite occasional complaints from commanding officers, his machinery did all that could reasonably be asked of it; and, considering the circumstances, even a bit more.

Isherwood's work as Engineer in Chief included more than the design of engines and the supervision of the men operating them. The Navy was unable to build enough ships to meet the sudden demands for blockade and river operations and, consequently, had to embark on a large purchasing program which actually obtained, by the end of the war, more than twice the number of vessels built by the government. The purchase of over 400 steamers, ranging from ferryboats to private yachts, was achieved largely through the remarkable efforts of George D. Morgan, brother-in-law of the Secretary of the Navy. Although accusations of nepotism and of excessive commissions plagued Morgan's attempts to procure this large fleet for the Navy, there now is little question that he performed invaluable service in obtaining so many vessels at unusually reasonable prices for the government. Isherwood, as the naval engineering expert in the department, was consulted frequently on prospective purchases, as the mechanical reliability of these vessels was paramount.

Perhaps the most time-consuming task for Isherwood was arranging for the building of the engines he had designed for government service. With only a rudimentary bureaucratic organization and a small staff, the executive officers in the Navy Department enjoyed a variety of duties, of which a most significant one was dealing with private contractors. Isherwood, in charge of steam machinery for the Navy, became responsible for making all the contracts for the construction of this machinery, as well as for procuring the necessary supplies, tools, and spare parts. Navy Yards were able to build some of the machinery, but the larger work had to be let out to private marine engine builders, most of whom were located in New York, Boston, or Philadelphia. In order to deal directly and effectively with these private builders, Isherwood spent much of his time traveling up and down the Atlantic Coast, negotiating for the Navy's machinery.

In contracting for machinery, Isherwood rarely relied on public advertisements for eliciting bids from contractors. Instead, he would announce that the government needed machinery of given specifications and would invite builders to state their quoted price for such equipment. Isherwood would, subsequently, send circular letters to engine builders throughout the country, asking them to contract for a pair of engines at a price based on the lowest estimate he had received. Since he would negotiate with any competent and reputable engine manufacturer, he often asked for contracts with builders who had declined to give him an initial estimate. That there were too few private companies to do all the work meant that Isherwood had to travel around the country trying to persuade engine manufacturers to take on extra business.

If his refusal to contract through open competitive bidding indicated favoritism, it was, nevertheless, a necessary practice. Machinery building was too important and too complicated to be trusted simply to the lowest bidder. To insure the best possible work, Isherwood had to limit his contracts to responsible builders and set a price where they could make a fair profit. Only by this procedure could he protect the quality of his machinery and discourage the horde of opportunists who plagued the government throughout the war with their impassioned and impractical offers to supply its needs.

The normal procedure after contracting with a private builder was for the government to make progress payments as the machinery moved towards completion. Since there was no contractual guarantee by the builder to achieve certain standards of performance, the government was not able to recover funds on machinery which later proved inadequate. Consequently, the only way the government could protect itself was to control the original specifications so rigorously that the contractor, in meeting them, would assure the reliability of the machinery. Isherwood thus prepared extremely elaborate and detailed specifications and drawings for builders, so that, as he expressed it, "there [was] not a bolt, a nut, or a screw left out." In this way no misunderstanding could occur, and no builder could find a loophole.[19]

Typical of Isherwood's negotiation procedure was a trip he took in August, 1861, to Philadelphia, New York, and Boston, in which he made arrangements with five of the largest and most experienced marine engine builders for the construction of 5 inclined steam engines. These engines were to be placed in paddle-wheel steamers built in the Philadelphia, New York, Charlestown, and Kittery Navy yards. In reporting about his trip to Welles, Isherwood took pride in having been able to make contracts for a building period 15 days shorter than that for similar engines previously built, and in setting a price 12 per cent less by weight or power.

Unfortunately, Isherwood usually had to give Welles less favorable news. Writing to him in February, 1862, the Engineer in Chief explained that work on the USS *Roanoake,* in process of conversion into an ironclad at the New York Navy Yard, had halted because he was unable to contract for a propeller shaft. All the private forges were taken up with other work, so that contractors had flatly refused to take on this necessary job.

Finally, Isherwood was able to persuade one builder to cast it in rough form, but when the unfinished shaft was taken to the government machine shop, Isherwood quickly discovered that the finished work could not be done there, since both the largest planer and lathe were inadequate for such a large piece of work and broke down when attempting it. The problem here was twofold; the Navy Department did not have sufficiently large tools to do the work, and private contractors, overwhelmed with both government and private construction, had neither room nor time to take on the work that the Navy yards were unable to handle.

Frequently, engine builders, greedy for profits or overly patriotic, took on Navy Department work they could not possibly handle. With their limited facilities they had to turn to other builders, subletting their contracts, which inevitably resulted in delays. The reputable firm of Pusey and Jones, in Wilmington, Delaware, contracted to do the machinery of the *Juniata,* promising to complete the work in 140 days. As they had no foundries and forges of their own, they had to sublet to another builder, resulting in a discouraging delay of 164 days more before the machinery could be delivered. Despite Isherwood's care in negotiating, time and again he discovered these "responsible" builders solemnly engaging for work which they alone could not possibly perform. Out on the Mississippi River or off the Carolina coasts, the Union Navy would ultimately suffer from the lack of a vessel.

There was one problem over which neither the contractors nor the Navy Department had control, and which threatened to throw the entire ship- and machinery-building industry into chaos. During the war the price of gold, measured in United States dollars, fluctuated greatly, reflecting a lack of confidence in the financial stability of the Union. Gold rose in price, especially after the issue in 1862 of greenback paper currency, resulting in a greater depreciation in the value of the dollar. The natural concomitant was soaring costs, particularly for materials and labor. Builders who had made what first had appeared to be profitable contracts with the government suddenly found that they would not make their profit and might even take a serious loss if they received only the contracted price for their machinery. Desperately, they turned to the Navy Department for extensions of time for construction, hoping also to avoid the penalty payments for delay which would further cut into their vanishing profits. Justifiably, they feared new contracts. The rate of inflation became so rapid that builders could no longer depend on short-term arrangements, let alone binding agreements for machinery which might take months to build.

Isherwood's task became more difficult as inflation cut into contractors' earnings. Writing to Assistant Secretary of the Navy Gustavus Fox, in August, 1862, the Engineer in Chief complained that he could get contracts for only two engines in Boston, because all the other builders objected to both the established price and the delivery date. Fearing that some of them might soon refuse to comply with contracts already signed, Isherwood explained that "prices of labor and material have risen enormously within a week." Part of the rise in wages was because of the scarcity of labor. To this initial shortage of skilled workers came the complicating factor of the draft, which indiscriminately seized men who were far more valuable to the government by remaining in their civilian occupation. Isherwood begged Fox to obtain for such workers a draft exemption which he hoped would not only enlarge the labor force, but also lower wage costs, thus allowing builders to accept contracts and fulfill them on time. "Indeed, if it is not done," warned the Engineer in Chief, "I am hopeless of the contracts already out being executed in any reasonable approximation to their time."[20]

To get his contracts accepted, Isherwood often had to increase the government's original price for machinery by 10 per cent and, in addition, extend completion dates for a month or two. "The most willing parties hesitate to accept," reported the discouraged engineer in another letter to Fox, "nothing will induce them but a price under which they will be safe at the expected advance." Because of the soaring labor costs, "common riveters for boiler work are now getting two dollars a day, and first-class workmen two and a half to three dollars...."[21]

Throughout the nation, shipyard workers, conscious of the sudden rise in their value, demanded better pay. "At the Morgan Works the men have struck for higher wages," Isherwood wrote to Fox, from New York city, "and at all the others the wages have been raised within the last week from 10 to 15 cents a day." Preparing to leave for Boston, Isherwood warned the Secretary of the difficulty in obtaining any contracts for engines. He would try to persuade the builders to take as many engines as possible, but this would be a formidable task, since, "the principal objection to the price offered is the expectation that labor and materials will advance 16 per centum before the work can be completed." Little wonder that the engine builders demanded a fat margin of profit.[22]

When the Navy Department planned to build a new group of larger geared engines, in 1863, Isherwood found only three builders willing to consider the job. Asking them to build the machinery in ten months, he found that the contractors refused to do the work in less than twelve. They had the facilities to do the job in ten months, they said, but lacked the raw materials and skilled labor, regardless of the price they were willing to pay to hire such labor. Isherwood had to agree to the longer time period and to the price of $700,000 for each engine, which was actually not too steep, he reflected in a letter to Welles, "when the present enormous rates for wages and materials are considered, with the great prospective increase anticipated in both."[23]

Traveling constantly to inspect vessels and make contracts for new ones, Isherwood could not afford any interruptions. When he fell from a ladder while inspecting the *Tacony*, in Philadelphia, he refused to stop work, although he had badly sprained his ankle and possibly fractured some of the smaller bones in his foot. With his ankle too swollen to determine the extent of his injuries, Isherwood wrote Assistant Secretary Fox on July 2, 1863, "I shall be in Washington Monday morning, on a pair of crutches I regret to add, but I shall be there." Furious at the delays caused by this ill-timed injury, Isherwood, oblivious of his own discomfort, apologized to Fox, "Nothing can be more inopportune than this accident. The pain and personal inconvenience I do not regard, but the loss of time, now so precious, grieves me much." Refusing to rest his ankle, Isherwood had it encased in a bandage filled with crushed ice, and hobbled into the offices of New York machinery builders, seeking contracts. His insistence on carrying on business as usual occasionally produced a ludicrous situation. When visiting the offices of Morris and Towne, New York engine builders, Isherwood was able to discuss engine contracts successfully with these men, but, as he wryly remarked, "with my back on the floor and Mr. Towne fomenting the ankle with ice water."[24]

As the war continued, the Navy Department found it harder to obtain its machinery. In February, 1864, when asking for estimates on a group of thirty engines for the *Sassacus* class of double-enders, Isherwood received only four proposals. Setting the price and building time on the basis of these estimates, he once again trudged around the country, successfully restraining his blunt and tactless nature while applying "much personal solicitation" to get twenty-seven of the engines contracted for. As ever, the price was too low and the time required was too short. In order to make his contracts at all, Isherwood had to promise not to exact penalty payments from contractors if they ran over the allotted time, so long as they continued to act in good faith. As he told Welles, he could not have contracted for more than six engines without this promise to make the forfeiture clause a dead letter.

"Unexpected and unprecedented rise of price in material and labor . . . disorganization of labor by strikes and its withdrawal for military purposes . . . difficulty of procuring materials . . ." appeared constantly in the reports of Isherwood to his superior, and there was no indication that conditions would improve.[25]

To further complicate the Engineer in Chief's duties, there was intense competition for scarce skilled labor between private builders and the government Navy yards. Handicapped by the inability to adjust their wage rates freely to meet the rising costs of labor, the Navy yards soon found themselves unable to compete with private yards for the inadequate supply of workers. As a result, the Navy Department had to farm out its repair work to private yards, "where the most exorbitant prices are charged," although tools lay idle in its own yards.

To alleviate this uneconomic situation, Isherwood urged Welles to obtain from the government an authorization to offer workers up to a 50 per cent increase in their wages, in order to keep them from leaving. Convinced that it would be more economical to pay much higher wages if this could keep the government yards busy, Isherwood asserted, in a letter to Welles, August 22, 1864, that the Navy would save an "immense amount of money annually by doing its own work." As it was, many costly tools were now standing idle while the government had to depend on the unpredictable, often irresponsible, and always costly private builders.[26]

Welles realized the need for utilizing government facilities to the fullest, and, along with his Engineer in Chief, he wished to increase greatly the size and capacity of the Navy yards. Urging the establishment of a new Navy yard for iron vessels and machinery, Welles followed Isherwood's and Lenthall's lead in seeking to lessen the government's dependence on private contractors.

During the war, Isherwood grew more and more pessimistic about the motives of private builders; and in his annual reports to Welles, he sharply criticized their business operations. Only interested in profits, they naturally preferred private to government work, he felt, since the latter, at best, provided a temporary boom in their industry, while they based their business over the years on private construction. As these builders operated with an eye to the least cost for themselves, they built engines for the Navy just to "answer a temporary purpose, using of course the poorest materials and least skilled labor because [it was] the cheapest." For this reason, engines built in government yards would always be better, Isherwood insisted, since the Navy Department stressed reliability and durability in the products it made for itself, regardless of cost. Not having to worry about profits "paid to wealthy capitalists," the government manufacturing facilities could afford to employ the most skilled workers at the highest rates, in order to insure the best workmanship.[27]

Isherwood felt that he had learned a valuable business lesson during the Civil War. Concerning the "popular impression" that the government could depend on private yards in time of war, he asserted, "such expectation would prove wholly fallacious." Drawing from his wartime experience with private contractors, he maintained that the facilities in private yards were inadequate even for the demands for privateers which, of course, would take precedence over any government work, as they had during the Civil War. Regardless of contract stipulations, private builders would always postpone government work to concentrate on the more lucrative private jobs. The Navy Department would always find itself saddled with inexcusable delays of vitally needed work while the private builders fattened themselves on immense profits, enough to pay, in one or two years, for the entire cost of equipping all the Navy yards. The only solution, Isherwood concluded, was for the Navy Department to have its own machine shops, large and complete enough to handle its own needs.[28]

Such growth of government facilities was not to occur in the 1860's. Apart from the strong objections naturally raised by private builders who had no desire to lose this government work, even if it was at times marginal, there was the thorny problem of where to build such facilities and how much they would cost. The bitter and lengthy debate in the late 1860's over the establishment of the League Island Navy Yard, at Philadelphia, is an instructive example of the problems that arose in locating a large new government facility. Moreover, after the Civil War, Congress was in no mood to authorize large expenditures on the Navy.

Isherwood's difficulties in finding skilled labor existed even in his own office. The increase of work brought on by wartime demands was more than his small staff could handle. In September, 1863, Isherwood asked Welles for another assistant draftsman, because of the great increase in the drafting department's work—most of which was on Isherwood's own engine designs. However, even with the Secretary's approval of this request, his problems were not over. A month later he wrote Welles, informing him that a naval engineer was currently filling in as the assistant draftsman because, at the government-regulated salary of $1,200, Isherwood could not find any civilian to take the job. As it was, he had raised the engineer's salary from $800 to $1,000 a year.

The Engineer in Chief not only struggled to find workers for his office, he had to strive hard to keep them there. Understandably, his young engineers chafed at their inability to be out on the firing line, but Isherwood turned a deaf ear to their requests for transfer. He detained them "without regard to . . . personal wishes or interest," Isherwood explained, because he didn't have the time to educate replacements; and once he had a trained, reliable man in his office, he was not about to let him go.[29]

The problems of the Navy were not always caused by the shortcomings of private contractors or the gyrations of the economy. If Isherwood indignantly complained about the unconscionable profit-seeking behavior of the builders, there were reasons enough for these men also to become impatient. The rapid advance in naval technology, spurred on by wartime demands, had its inevitable repercussions as builders struggled with new and ever changing designs. Much of the delay in building was the result of more than constructors' procrastination; the fertile minds of both civilian and naval inventors made their contribution to the chaos. Assistant Secretary Fox, in a letter to Alban Stimers, dated February 25, 1864, indicated this influence on mounting costs and delays as he attacked "those horrible bills for additions and improvements and everlasting alterations, all of which have cursed our cause and our Department." Doubling the contract price with such alterations, additions, and improvements was common, as Isherwood and Lenthall, supervising the design and the building of vessels, continually disrupted the progress of construction with new specifications and changed designs.

As the Civil War progressed, it became obvious that certain of the Navy Department bureaus were overloaded with work. In particular, the Bureau of Construction, Equipment, and Repairs was clogged with the excessive responsibilities for both the vessels and machinery of an expanding fleet. Steam engineering had suddenly become paramount, and expenditures in this area were soon to exceed any other in the Navy Department. "Steam has become such an indispensable element in naval warfare," Gideon Welles reported, March 25, 1862, in a letter to the House Committee on Naval Affairs, "that naval vessels propelled by sails only, are considered useless for war purposes." The last sailing vessel for the

Navy, the *Constellation,* had been completed in 1855, and soon the Navy would be "exclusively a steam navy," and a very large one.[30]

To meet the demands for a more efficient department, Welles combined with Senator James W. Grimes, of Iowa, in early 1862, to plan a Navy reorganization which would create three new bureaus, while removing the excessive work load from the existing ones. Their plan, which became law on July 5, 1862, broke apart Lenthall's bureau, leaving him the function of construction and repair, placing the equipment duties into a new Bureau of Equipment and Recruiting, and, in particular, creating a new Bureau of Steam Engineering. The tenure of bureau chiefs changed from an indefinite status, subject to the pleasure of the President, to a regular four-year term. Salary would be $3,500, in lieu of regular Navy pay for those bureau chiefs who were naval officers.

The new steam engineering bureau was to have six clerical assistants to the chief—two clerks, two draftsmen, a messenger, and a laborer—with salaries high enough, it was hoped, to attract civilians to fill the positions. Beating down an attempt to open up the position of bureau chief to civilians, Grimes managed to restrict the position to chief engineers in the Navy, a move which was greeted with enthusiasm by the corps of naval engineers, but met with a civilian reaction which presaged trouble for the future.

Many naval engineers, confused by the new structure of the department, thought that the chief of the bureau and the Engineer in Chief would be separate positions, requiring two men. Several of the senior chief engineers thus eyed the new office with interest, quietly marshaling support for their cause, while publicly disclaiming their own ambition. This was not what the Secretary of the Navy had intended. In creating the new bureau, Welles had expected the Engineer in Chief simply to switch hats, continuing his former duties without interruption. On Welles's advice, President Lincoln nominated Isherwood to be the new bureau chief, on July 11, 1862. At the same time, he nominated John Lenthall to continue his work as chief of the Bureau of Construction and Repair.

The nominations went routinely to the Senate Committee on Naval Affairs, but six days later Isherwood and Lenthall realized they might be in trouble as their nominations were tabled and the Committee on Naval Affairs was discharged from further consideration of the matter. Although the Senate had not confirmed the presidential nomination, Welles sent Isherwood his commission as bureau chief as of July 23. The Engineer in Chief gratefully accepted it, giving Welles his thanks "for this continued evidence of the great confidence and trust you have reposed in me; and to assure you that no efforts will be wanting on my part to justify your selection."[31] In a situation like this, the support of his superior was no small matter.

As the months went by, Isherwood served as bureau chief, but there was no further move in the Senate to act on his nomination. Finally, on December 1, Lincoln again nominated Isherwood and Lenthall as bureau chiefs. After the nominations went to the Senate Committee on Naval Affairs, there was a period of bitter debate concerning the merits of the two men. On December 22, James W. Grimes presented to the Senate three "memorials of proprietors of the principal marine-engine building establishments in the country," along with one from engineers in the Navy, all asking for the confirmation of Isherwood as bureau chief.[32] At the same time there was fierce opposition, as disgruntled contractors, administration opponents, and professional rivals sought the removal of both Isherwood and Lenthall. By the end of January there was still no resolution of the issue. Welles wrote to his son Edgar that the Senate refused to confirm the two men and that he "would not be surprised if matters go hard with Isherwood," because of the sharp criticism of the engines he had designed and had placed in government vessels.[33]

There were many influential line officers who, despite their absence from Washington, could wield powerful influence against both Isherwood and Lenthall. Reflecting their attitude was a colorful diatribe by David D. Porter, which, though written during the previous year, indicated the distrust and contempt felt by line officers for the bureau chiefs.

"That man Lenthall," fumed Porter, "has been an incubus upon the Navy for the last ten years." Convinced that the naval constructor was the tool of Stephen Mallory, former chairman of the Senate Naval Affairs Committee and now secretary of the Confederate Navy, Porter claimed that Lenthall and Mallory had plotted to "throw as much of the Navy as possible into the hands of traitors" in early 1861. Lenthall was responsible for keeping the *Merrimack* at Norfolk, Porter asserted in a letter to Gustavus Fox, July 5, 1861; for as a civilian, he tried to get the Navy yards out of the control of naval officers and into the hands of civilians, so that at the proper time he could hand over the yards to the South.[34]

Declining to assail the Engineer in Chief as a traitor, Porter contented himself in the same letter to Gustavus Fox, with a contemptuous description of "that little fellow Isherwood who will take all the signs in Algebra to prove how many ten penny nails it will take to shingle a bird's nest, who will bring out more equations to prove that a pound of water can be so expanded that it will make a ship go 25 miles an hour, and yet he can't make an Engine." With a convenient inaccuracy of recollection, Porter related that Isherwood had been "Engineer with me 9 months; I took him out of the Engine Room for incapacity, [but] he may have improved since...."[35]

Porter, in his unique way, mirrored the line officers' frustration with a Navy Department that, to them, seemed unable to meet their needs and appeared to disregard or depreciate their views. In the midst of battle, they resented those men back in Washington who appeared indifferent to their problems which, the officers insisted, should be of paramount importance in the conduct of the war. The never-ending accumulation of data for Isherwood, continuing on vessels even as they came under fire, drove commanding officers to distraction. They had no patience for work which seemed to bear no relation to the actual operation of their vessels. Isherwood's only duty and responsibility, they believed, was to produce engines which would never break down and engineers who would never make mistakes. As neither of these demands would ever be wholly met, Isherwood was bound to become a target of frustrated and often vengeful men who had become convinced that he neither knew nor cared about their vital interests out on the firing line.

Despite the resistance to his nomination from so many quarters, Isherwood emerged triumphant. When President Lincoln nominated him for the third time on March 7, 1863, the opposition collapsed; and three days later, the Senate considered and confirmed Isherwood's appointment. By an order of the Secretary of the Navy, dated March 13, Isherwood, as bureau chief, now ranked with commodores, taking precedence among the other staff officer bureau chiefs according to the date of his commission as a naval officer. Lenthall had met less opposition and, consequently, received confirmation as chief of the Bureau of Construction and Repair on February 21, 1863.

As Engineer in Chief under Lenthall, and then as Chief of the Bureau of Steam Engineering, Isherwood, throughout the Civil War, mastered a tremendously complex, arduous, and often frustrating assignment. In later years he would describe the full scope of his duties with a succinct and disarmingly direct statement:

> During the war the appointing and detailing of nearly two thousand Engineers were in my hands, besides the designing of the machinery of several hundred naval vessels, and the direction of the repairs and alterations of as many more. The contracts for all these, and for the immense quantity of engineering supplies, were part of my duty, and in addition there was the examination of the innumerable plans of vessels and machinery daily presented to the Department, and the writing of complete reports upon the same.
>
> In short, everything connected with the engineering of the Navy during the war—in the widest sense of the word—was under my immediate direction, and I was held responsible for it. The nominal and the real responsibility were the same, and no Boards of officers were allowed to either shield or assist me.[36]

Characteristically, he had preferred to depend on his own judgment, rarely asking for help as he brought the vast, unco-ordinated aspects of wartime naval engineering together into a unified whole. Absorbed in his work and convinced of its importance, he needed no praise nor prodding. His country had called him to this service; he could do no less than his best.

# *III. Isherwood and the Ironclads*

Naval technology received a tremendous impetus from the Civil War. As in any period of rapid advance, however, there was continual disagreement and resistance to innovation. Not only was there heated debate over the designs and materials to be used in naval construction, but there also emerged a fundamental divergence of opinion over the correct strategic use of warships, and, consequently, over the type of warship which the Navy should employ.

The ironclad warship was not a Civil War innovation. The war did provide, however, the first practical demonstrations of these armored vessels in action against both wooden ships and other ironclads. From these experiences grew a number of conclusions which served to shape American naval strategy and tactics for the next generation.

The most striking American contribution to armored ship design was the monitor concept of John Ericsson. This temperamental genius, reluctantly offering his services to a heretofore ungrateful nation, produced a unique revolving-turret vessel which, by its dramatic encounter with the Confederate ironclad *Virginia* (formerly *Merrimack*) in March, 1862, captured the imagination of much of the Navy Department and won the spontaneous and unqualified devotion of the Union. The monitor idea, however, was not the only ironclad plan offered during the Civil War, and in light of future American naval developments, it was not the best.[1]

American interest in armored warships predated Ericsson's offer to build the *Monitor*. In July, 1861, New Hampshire-born Senator James W. Grimes, whose present residence in landlocked Iowa did not preclude a continued fascination for naval affairs, brought before Congress a bill for the construction of armored ships. Having read everything written on the subject during the last two or three years, Grimes was thoroughly acquainted with the French and English experiments in the field. Despite the fact that both of these countries had produced seagoing ironclads (the armor-plated, wooden-hulled *Gloire* of the French, and the larger iron-hulled and armored

*Warrior* of the British), Grimes believed that the only indisputable value of ironclads was for the defense of harbors, this being "the opinion of the scientific men in this country and in Europe," according to the Senator. On this basis, he recommended the construction of floating armored batteries for harbor defense, in case a war with foreign powers developed from the present conflict with the South.

Within the Navy Department, there was little initial enthusiasm for ironclads, at least for the type Grimes proposed. When the English shipbuilding firm of John Laird Sons and Company offered, in the summer of 1861, to build armored gunboats for the Union, they were turned down by Gideon Welles. However, Assistant Secretary Gustavus Fox felt differently:

In '61 before Errickson [*sic*] gave us such an admirable inshore impregnable battery we had nothing, nor the offer of anything. Isherwood thought ironclads a humbug and Lenthal[l] shrank from touching the subject first at that period when fatal days and months were passing.[2]

Fox, convinced that the shipbuilding experts in the department were dragging their heels, encouraged the Lairds, since "no one at the Dept. objected to receiving plans, specifications and offers from any part of the world." Not expecting the Navy to order English-built warships, Fox instead saw these offers as a way to obtain "something tangible to force Lenthal[l] to work out our great want."[3]

If John Lenthall, designer of so many graceful wooden-hulled steam frigates, shrank at the thought of sheathing them with tons of iron plate, or even cutting them down into squat, ugly floating batteries, stripped of masts, sails, and dignity, he did not do so for very long. In October and November, 1861, with the assistance of his Engineer in Chief, he produced a plan for an iron-plated steam battery to be built in five and one-half months at a cost of $530,000.

This twin-screw, wooden-hulled vessel featured two cylindrical towers, or turrets, similar to Ericsson's, but built on a system designed by the English Captain Cowper Phipps Coles. Instead of mounting the turret on a central spindle, as Ericsson proposed, Lenthall and Isherwood chose to have it rest on bearings which ran around its circumference. The base would be protected by a glacis plate, to avoid the vulnerability of Ericsson's exposed turret base. Large single iron plates, between three and one-fourth and four and one-fourth inches thick, would cover the hull, instead of weaker, laminated, thin armor plates. The turrets would have the protection of five inches of armor, backed by ten inches of oak.

The appearance of a department proposal for twenty of these ironclads threw Ericsson and his supporters into consternation, since they had hoped for further contracts for vessels based on his *Monitor* principle. Ericsson attempted to convince Welles that his turret design was earlier and better than Coles's "abortive scheme," while his associates assured Isherwood that the Ericsson turret would be best for the department ironclad. Ericsson refused to write to Isherwood directly, confiding to Fox, in a letter dated March 14, 1862, that he preferred to keep "my own counsel on account of the most unexpected march which Mr. Isherwood stole on me...." Isherwood not only had chosen a foreign design in preference to Ericsson's, but had planned his own turret engines, which Ericsson grumpily accused of having too much power.[4]

As it turned out, Ericsson had little to worry about. While the Senate delayed in appropriating money for the department's ironclads, Ericsson's supporters brought heavy pressure on congressmen to favor the monitor design. Difficulties in obtaining the heavy iron plates forced the government to seek its armor abroad, further delaying the start of construction. Finally, the success of the *Monitor* at Hampton Roads brought Fox solidly over to Ericsson's side, and induced the government to order twelve of his monitors.

Although their ironclad project had failed this time, Isherwood and Lenthall were finally able to persuade the Navy Department to build monitors from their own plans, but even here the department rejected the Coles turret in favor of one using Ericsson's central spindle. In 1862 four of these large, double-turreted ironclads were placed under construction in Navy yards, two of which, the *Miantonomah* and *Tonawanda,* were to have Isherwood's engines, while the other two used Ericsson's machinery. These vessels were quite successful, and two of them later displayed the ability of monitors to cruise at sea. In November, 1865, the *Monadnock,* with Ericsson's engines, made the arduous trip around Cape Horn, arriving at San Francisco the following June. Later in 1866, the *Miantonomah* vindicated Isherwood's machinery design by crossing the Atlantic, carrying Gustavus Fox to England and Russia.

These low-freeboard, turreted steam batteries for coast and harbor defense were not Isherwood's and Lenthall's major interest. Far from ignoring the potentialities of armored warships, they recognized a far greater role for such vessels. On March 17, 1862, a little more than a week after the encounter between the *Monitor* and the Confederate *Virginia,* Isherwood and Lenthall sent a long memorandum to Welles in which they formally stated their views.

First of all, they insisted, the government should build its own facilities for making the heavy iron plating needed for the hulls and armor of warships. This factory should be a civilian establishment, fabricating and warehousing specialized iron products, rather than being simply another Navy yard. The need for such a factory was evident in the failure of the government to get sufficient armor from private manufacturers who lacked either the proper machine tools or the ability to make heavy iron plates. In addition, private operators were unwilling to take on work of such size when there was no assurance of future continuous employment of the specialized machinery they would have to obtain to manufacture iron plates.

For Lenthall and Isherwood, there could be no doubt that large iron-working facilities were necessary, because these were required to satisfy America's need for a large, armored Navy. Since the security of the United States, threatened by the transoceanic capabilities of great foreign powers, depended on naval prowess, "the necessity and importance of an establishment that is to provide a future navy sufficient for securing a country like ours from foreign aggression," they declared, "is . . . a national question, second to none. . . ."[5]

With a new government factory for iron plates and with the present upheaval in naval technology, Isherwood and Lenthall believed the United States could "start equal with the first powers of the world in a new race for the supremacy of the ocean." With no vested interest in obsolete workshops or in a large Navy, the United States would actually have a considerable advantage. Profiting from the experience of others without any great expense of its own, it would soon be able to produce "a fleet of first-class, invincible ocean ships."[6]

Here lay the key to the argument of the engineer and naval constructor. They discarded the currently popular strategic concept of defense, based on the low-freeboard, single-turreted monitor which operated only in harbors, bays, and river mouths. Instead, they proposed a sweeping revision of American naval policy, declaring, "this country must hereafter maintain not only a larger navy than it has heretofore done, but of an essentially different character...." This difference would lie in the construction of "cruising vessels on which alone reliance must be placed for offensive war." These ships, specifically designed for offense rather than for inshore defense, would be "frigate-built, iron steamships of sufficient strength to be used as rams, clad with invulnerable armor plates, furnished with maximum steam power, and of a size larger than any vessel we now possess." Such large, extremely fast vessels should be able to carry the heaviest guns, they believed, in order to engage the most formidable seagoing ships the enemy could offer.[7]

Isherwood and Lenthall, in calling for the fastest, biggest, and most heavily armed and armored ships, were doubtless asking for too much, since the qualities they specified could never all be combined in a single ship, except at enormous cost and by involving tremendous size. At best, some of the qualities, such as "invulnerable" armor, would have to be compromised to attain others, such as speed. In addition, it is doubtful if American technology was sufficiently developed at that time to produce such an immense and complex vessel.

The most significant element in their argument, however, was neither impractical nor visionary. It was the strategic doctrine on which they based the need for and use of such ships:

It is obviously cheaper, more effective, and more sustaining of the national honor to preserve our coasts from the presence of an enemy's naval force by keeping the command of the open sea, with all the power it gives of aggression upon his own shores and commerce, than to rely on any system of harbor defence which requires every point to be protected that may be assailed by any enemy, having, in that case, the choice of time and place, and the advantage of perfect security for his own ports and commerce.[8]

In this one sentence, Isherwood and Lenthall revealed the fatal weakness in relying upon monitors and presented a view of naval strategy which, thirty years later, would evolve, in the hands of Alfred Thayer Mahan, into virtual gospel for the naval powers of the world.

Command of the sea was also essential for the success of a traditional American naval institution, the privateer. Correctly reading the real naval lesson for America from the War of 1812, Isherwood and Lenthall stated, "a clear coast is manifestly essential to any effective system of privateering." All the harbor defense monitors in the world would not help the American commerce raider, they argued, for although the enemy might be kept out of a port, the monitors "could not drive him from its gates; and if blockaded by his large iron-plated steamships, no privateer could either get out himself, or send in a prize."[9]

For this reason, they decided, monitors might be "valuable adjuncts," but they could never "constitute a navy, or perform its proper functions."[10]

In a striking conclusion to their letter, the Engineer in Chief and the Chief of the Bureau of Construction, Equipment, and Repairs drew on their clear, strategic vision and their command of stirring rhetoric to declare:

Wealth, victory, and empire are to those who command the ocean, the tollgate as well as the highway of nations, and if ever assailed by a powerful maritime foe, we shall find to our prosperity, if ready, how much better it is to fight at the threshold than upon the hearthstone.[11]

Isherwood and Lenthall, in this forcible expression of their views, predated Mahan in their recognition of "command of the sea," but it would be incorrect to say that they had developed as thoroughgoing and modern a doctrine as he did during the 1890's. They still clung to the importance of the privateer, and their seagoing ironclad was a superfrigate whose commerce-destroying function took precedence over that of battle with enemy warships. Their cruiser would conduct an offensive war, but this meant depredations upon enemy commerce and coastlines. They did not, at any time, argue for the use of ironclads in squadrons, particularly for the purpose of doing battle with the enemy's assembled fleet. They relied on the tradition of the *Constitution,* a ship strong enough to engage all but the heaviest of the enemy's warships, but a ship which, basically, operated alone and usually as a commerce destroyer.

In the description of their seagoing ironclad, they made no mention of cruising under sail, but given the current state of technology, they would have been forced to rely largely on sail power, since such a ship could carry only enough coal to permit short bursts of steaming at maximum power. The inefficiency of steam engines of the early 1860's, coupled with the lack of foreign coaling bases for an American vessel, necessitated reliance on sail power for most of any cruise. Their vessel was thus truly a concept of the 1860's, embodying all the limitations of that period. Its importance as a ram—a tribute to the success of the Confederate ironclad at Hampton Roads—and its proposed use as a single commerce destroyer, rather than as a component of a modern battle fleet, shows that both technology and strategic concepts had to mature before a doctrine of modern sea power could emerge.

The letter of Isherwood and Lenthall did impress at least one important individual: the man to whom it was addressed. In his annual report for 1862, Gideon Welles asked for a government yard and depot for an ironclad Navy, remarking that in March and in June he had presented this request to both naval committees of Congress. In addition, Welles asked for a "formidable navy," not just of light vessels for coast defense, but of warships big enough to have great speed and the ability to "seek and meet an enemy on the ocean." To achieve this capability, the Secretary recognized that American warships would have to be extremely large in order "to obtain the enormous steam power essential to great speed." By 1862, Welles considered the ironclad to be the inevitable standard for warship construction, and for this reason he reiterated Isherwood's and Lenthall's demand for government manufacturing facilities to handle the specialized requirements for ironclads.[12]

In his annual report for the following year the Secretary amplified his argument, insisting that "to maintain our rightful maritime position, and for predominance upon the ocean, vessels of greater size than any turreted vessel yet completed may be essential." He demanded big guns, heavy armor, and the capability for extended cruising, so that American ironclad cruisers would have "all possible strength, endurance, and speed."[13]

The first chance for Isherwood and Lenthall to build a seagoing ironclad came quickly. On March 19, 1862, recognizing the difficulties in obtaining congressional acceptance of their entire program, they submitted a modified plan by which they would take existing wooden steamers and convert them into ironclads. They proposed to take the large screw frigate *Roanoake,* sister ship of the *Merrimack,* cut her down to the gun deck, and then plate her with armor four and one-half inches thick on the sides, three inches on the ends, and two and one-fourth inches on the decks. They would strengthen the bow so the vessel could be used as a ram "for cutting down wooden vessels," as the *Virginia* had done so successfully to the *Cumberland* at Hampton Roads. The converted *Roanoake* would be armed with eight 12- or 15-inch guns mounted in four Coles-type turrets, which would be arranged along the center line of the vessel. The estimated cost for conversion was $70,000 for the hull, $25,000 for the additional machinery, $190,000 for the Coles "towers," and $210,000 for armor, a total of slightly under half a million dollars. They initially estimated the time for conversion to be three and one-half months, although the job could be done in sixty days if not for problems in obtaining iron plates.[14]

Construction delays quickly appeared. By insisting on the heavy four-and-one-half-inch armor instead of laminating several one-and-one-fourth-inch plates, Isherwood and Lenthall had to extend the conversion time to over five months. As no Navy yard could produce such heavy armor, the government had to go to a private contractor, the Novelty Iron Works, in New York city, both to get the armor and to build the turrets. The real difficulty, for once, proved to be not the contractor, but the design of the vessel. Isherwood and Lenthall soon discovered that the ship could not support four turrets. Forced to modify their plans to include only three, and those of Ericsson's design, they still ran into trouble. The immense weight of the turrets and the armor proved too great for a vessel of that size. The converted *Roanoake* had far too great a draft and proved unseaworthy.

Although a failure as a practicable warship, the *Roanoake* embodied a number of design concepts which would come to full fruition in the twentieth century. By cutting down the frigate and removing her masts and yards, the department created a high-freeboard, seagoing ironclad completely dependent upon steam power. She was the first ship with more than two turrets on a center line. The idea of an all-big-gun, fully armored vessel with the turrets set in a line along the keel would ultimately become the orthodox design for modern battleships of the twentieth century. The famous British *Dreadnought,* launched in 1905 as the first of the all-big-gun battleships, was thus a direct descendant of the *Roanoake.* In both cases, the ships were designed for high speed, while carrying formidable armament in armored turrets. The *Roanoake* may have been an abortive design for the period, but she heralded to an unconscious naval world the shape of things to come.

The success of the *Monitor* had helped to convince Welles of the desirability and practicability of a fleet of ironclads for the Union Navy. His board of line officers, which met in 1861 to choose plans for the first ironclads, had been so diffident about its task that when Welles decided to establish another board in early 1862, he called instead upon his experts in naval architecture and engineering. On March 26 he appointed a new group, consisting of Commodore Joseph Smith, chief of the Bureau of Yards and Docks, Lenthall, Isherwood, Edward Hartt, a noted naval constructor, and Daniel Martin, former engineer in chief of the Navy. Those experts were ordered to examine plans and specifications submitted for harbor-, river-, and coast-defense ironclads. Meeting in Lenthall's office, the board started work on March 28, and made its first report to Welles on April 9, 1862.

The ironclad board first considered plans for riverboats, and then spent a month on the designs for four double-turreted iron monitors to be used for river and coast service. Almost immediately there was disagreement among the officers. The majority, comprised of the engineers and naval constructors, were "decidedly in favor of building all ships of war of iron," but not so the chairman, Commodore Joseph Smith. This old line officer was in no hurry to accept the latest advances in warship design, especially when there was no assurance that the newest plans were the best. He urged that ironclads still be constructed of iron and wood, noting that builders were not yet skillful and sufficiently experienced with iron vessels, and that in England naval constructors were currently using wooden hulls for ironclads. In addition, the majority of the board, Smith felt, had not acted fairly towards the various proposals they had examined. Their minds were closed to outside suggestions, and they insisted on recommending only plans identical to those of Isherwood and Lenthall.[15]

The naval constructors and engineers, led by Isherwood and Lenthall, defended their position in a letter to Welles written several days after Smith had submitted his minority view. Rather than the monitor-type of harbor-defense ironclad, with its iron hull, thin laminated plates, single screw, and single turret, they preferred a department plan which had been devised by Isherwood and Lenthall. Their ironclad would have thick single plates, twin screws for greater maneuverability, and two "revolving towers." Admitting that they originally had chosen wooden hulls for ironclads, because of their speed of construction and conversion, the majority said that now all newly built ships should be constructed entirely of iron. Moreover, laminated armor allowed corrosion and was not as resistant as a single, thick plate. To obtain greater fire power and flexibility, they preferred two turrets instead of one for the river and harbor ironclads. They also explained that, for such restricted waters, guns in revolving turrets would be superior to those in casemates because of their ability to cover a field of fire with greater speed and precision. For sea service, however, the engineers and constructors would not approve a monitor design because of the potential problems involved in working a turret in a rough sea.

The real concern of the majority was to determine whose plans should be used. Instead of accepting a number of varying designs submitted by private builders, they preferred to have the Navy Department prepare its own plans and specifications and then advertise for bids from constructors. They argued that naval experts best knew what was required of an ironclad and could, therefore, design the most practicable vessel. In addition, the use of the department plan would assure a uniformity of vessel type, making construction easier and cheaper, and affording less risk for the private contractor. Despite their intention to promote Isherwood's and Lenthall's design, the majority deigned to recognize briefly the contributions of John Ericsson, saying that, in an emergency, the Navy might utilize the improved single-turreted *Monitor* design, since this type of vessel had undeniably given satisfactory results.

However, they quickly added, neither Ericsson's monitor nor the department's harbor-defense ironclad would be able to protect any port against a large, seagoing ironclad. Shore batteries and channel obstructions would have to supplement any coast-defense ironclad to save American ports from large enemy warships. Although the majority declined at this time to offer a design for a seagoing ironclad, they strongly implied that the Secretary should recognize the necessity, so persuasively argued by Isherwood and Lenthall, for such a cruiser, because only with protection from this type of vessel could American harbors ever be really free from attack.[16]

The Engineer in Chief and the Chief of the Bureau of Construction, Equipment, and Repairs, after their letter to Welles on naval policy, wasted no time in presenting to him plans for a seagoing ironclad. While converting the *Roanoake* and reviewing the ironclad proposals submitted by civilian designers, they were busy preparing plans for a ship which would meet their own requirements. By October, 1862, they had finished their work, and ran a public advertisement on October 30, asking for bids on a 7,300-ton "iron-clad steamer."[17]

This vessel, Isherwood and Lenthall decided, should not have turrets because of the difficulties in operating them in a rough sea. Instead, the guns should be mounted in a casemate—a long, armored, rectangular box set on the vessel's hull. With a superstructure similar to that of the Confederate *Virginia* or the highly successful Union ironclad *New Ironsides,* their seagoing cruiser would have most of its guns mounted broadside, so that the fore-and-aft fire would be relatively weak. Unlike the *Virginia* or *New Ironsides,* however, the department plan required great speed and a hull shaped for easy handling at sea.

Not only would this seagoing cruiser have an iron hull, but its internal beams would also be of iron, lending greater rigidity to the structure so that it could be used as a ram. Its armor would be of thick plates, rather than the thin, laminated ones so frowned upon by Isherwood and Lenthall. The vessel would be heavily armed with ten 15-inch guns; and with its spar deck 7 feet, 9 inches above the water line, the warship would be able to keep its guns out of water in all but the heaviest of seas. Power was to be supplied by two Isherwood-designed, geared, horizontal engines, with large cylinders of 108-inch diameter and with a 6-foot piston stroke. All the machinery was to lie below the berth deck and between the fore-and-aft bulkheads of the ship, thus keeping it protected from enemy fire by being below the water line.

The department plan drew immediate criticism. *The New York Times,* a newspaper notoriously hostile to the Navy Department, published an editorial, on December 25, in which it coldly examined the ironclad cruiser in detail, while saving its rancor for the Engineer in Chief. The estimated cost of $4,200,000 was excessive, *The Times* felt, and to make things worse, the design was basically faulty. Using a thin skin of armor all over the vessel instead of protecting its vital spots by heavy layers was absurd, since, with only four and one-half inches of armor, none of the ship would be safe. The choice of single armor plates was poor, especially since Isherwood and Lenthall did not use any backing. England and France, *The Times* pointed out, were using heavy wood backing for their armor at this time, and the Ericsson-designed *Dictator* class of large monitors not only used fifty inches of oak backing, but its armor was ten and one-half inches thick. Modern ordnance could easily penetrate the puny four and one-half inches allowed by Isherwood and Lenthall; and at this very moment, one of their fellow bureau chiefs, John Dahlgren, was successfully firing eleven-inch shot through these same four and one-half inch plates.

The machinery was particularly bad, declared *The Times,* in the same editorial, because it was designed by the man who had "neutralized the entire engineering ability of the country" by selfishly building most of the government engines from his own designs. How could any machinery be worthwhile, the paper inquired, if it had sprung from the eccentric brain of Isherwood? This was the man who advanced spurious theories of steam engineering and who disregarded everything in a vessel but the machinery, often placing it above the water line through his negligence of the fundamentals of warship design.

Professional hostility towards Isherwood, fanned by such editorials and articles in the press, forced Welles to conduct an investigation of Isherwood's engines. On January 9, 1863 he appointed a board of five civilian engineers to examine the government designed machinery. Not surprisingly, when the board reported to Welles on February 18, they were highly critical of the Engineer in Chief for so successfully retaining control over the engine designs and construction for naval vessels. Referring to the seagoing cruiser, the board were extremely skeptical of its feasibility. Noting that it would have a twenty-foot draft and would displace eleven-thousand tons, the board doubted if it could approach its projected speed of sixteen knots. With its "inferior" boilers, designed by Daniel Martin, and its geared engines, they felt the ship would be fortunate to make over twelve knots. Other engineers, especially civilians, could have done everything better, the board of civilian engineers assured Welles.[18]

In an attempt to answer such criticisms, Isherwood and Lenthall brought out a revised ironclad proposal in March, 1863. Refusing to compromise on size, they designed this vessel to be eight-thousand tons burden with a draft of twenty-one feet, and to carry ten guns in casemate, each weighing twenty-five tons without its carriage. However, Isherwood significantly altered the machinery. He dropped his geared-engine idea and substituted two pairs of direct-acting, horizontal engines, in which the engine and propeller shaft would work in a one-to-one ratio. In addition, he required a sixteen-hundred-ton coal capacity for this "iron iron-clad sea steamer" in order to assure a reasonable cruising range.

Still there were problems. The size of the proposed vessel intimidated builders who already had sufficient problems trying to complete much smaller vessels for the department in the face of rising costs and the lack of facilities. Consequently, it was not until the end of the year that private builders began tentative negotiations with the department concerning a large, seagoing ironclad. On December 5, Merrick and Sons, of Philadelphia, builders of the *New Ironsides,* wrote to Lenthall about the possibility of building another ironclad. They estimated it would displace 8,603 tons with its iron hull and armored casemate, 2 pairs of engines, 4 guns, and a crew of 500. Encouraged by Lenthall's interest in such a ship, Merrick and Sons came back, on December 28, presenting him a formal offer to build a 7,100-ton vessel. This ship was to be 470 feet long, 66 feet wide, and would draw 20 feet. As a single-decked steamer, carrying 4 guns in casemate, the ironclad was to attain a speed of 16 knots for 24 consecutive hours in smooth water. The builders chose 2 pairs of vertical direct-acting engines for power and planned to cover the hull with 4½ inches of armor, while the casemate would be protected by 2 layers of iron plates totalling 5 inches. The cost of this ship was to be $4,300,000, and it would take 3 years to build.

These were formidable obstacles to its actual construction. Although Isherwood and Lenthall enthusiastically pushed their program for large, seagoing ironclads, Gideon Welles, with an eye on cost and availability, was in a quandary. Admitting that large, expensive vessels were not needed now for actual battle against Southern steamers, he saw, on the other hand, that such a class of warship would be a powerful deterrent to England and France at this time when their sympathies were strongly and openly on the side of the Confederacy. In one way, Welles argued, large vessels would be economical, since they would best serve as a deterrent; yet in peacetime, there would be only a few, thus avoiding a heavy commitment in manpower and maintenance funds. The trouble was that the Navy was in serious need of blockading vessels at the moment. It was not fighting a large Navy, but only fast blockade-runners, and large ironclads were not practical for such operations. In addition, the monitor enthusiasts were steadily growing in power and influence, especially with the launching of the powerful *Dictator*, Ericsson's monitor; and the department could commit itself to building only a few large ironclads at any time.

While Welles withheld support of the large seagoing ironclads, offers began to come into the department in answer to still another large vessel proposed by Isherwood and Lenthall. This ship was to measure 475 by 63½ feet, with almost a 24-foot draft, and would be an iron-hulled, armored, casemate battery of 7,000 tons burden. The engines, smaller than earlier models, would consist of 4 cylinders, of 90-inch diameter and a 4-foot stroke; but this time they would use tubular boilers with a greater heating capacity than those previously proposed.

Wary of continued inflation and disillusioned by previous experience in shipbuilding for the government, private constructors submitted bids which only confirmed Welles's fears of excessive cost and time in building. Reany, Son and Archbold, of Chester, Pennsylvania, first to send in an offer, required a full three years for construction, at a cost of $5,500,000. The Atlantic Works, in East Boston, Massachusetts, needed as much time and asked for $200,000 more. In New York the Continental Works promised to take only thirty months for construction, but demanded a price of $5,950,000, while the Fulton Foundry, in Jersey City, had to have thirty-three months to turn out the ironclad for the staggering price of $6,948,069. A commitment of so much money and so much time was more than the government could afford. Isherwood and Lenthall once again saw the plans for their great ship relegated to the files.

While the two engineers were attempting without success to build a large, casemated ironclad, the Assistant Secretary of the Navy was busy in a different direction. Gustavus Fox, ever since the success of Ericsson's *Monitor* at Hampton Roads, had been convinced that the monitor design was superior to any other. Knowing he could find no support from the two bureau chiefs for a seagoing, turreted ironclad, Fox determined to seek outside assistance. He soon found a warm advocate for his views in James B. Eads, the able and energetic St. Louis contractor who had already built a fleet of successful ironclad gunboats for Mississippi River operations. Fox first sounded out Eads in October, 1863, asking him to design plans for an immense ocean steamer which would carry four formidable guns in a massive turret protected by eighteen inches of armor. Roughly the same size as Isherwood's and Lenthall's seagoing cruiser, Fox's ironclad was to steam at eighteen knots, despite the weight of the turret and the twelve-inch side armor he required. This awesome warship should be capable of ramming, Fox told Eads, October 23, 1863; and its armor and armament should make it able to command the seas, even into the English Channel.

Eads responded enthusiastically to Fox's plan for a "monarch of the seas and the bulwark of the nation." He suggested, in a letter to Fox dated November 4, 1863, that the four guns be increased in size to twenty inches in diameter, and he envisioned a turret thirty-one feet in diameter to hold them. After spending several months on the design, he reported to the Assistant Secretary that he could build the vessel with solid iron armor eleven and one-fourth inches thick at the water line, extending for two hundred feet. Despite the weight of all this iron plate, plus that of the turret and guns, his ship would weigh no more than the casemate ironclad of Isherwood and Lenthall. Eads furthermore promised that firing the guns in a heavy sea would not be a problem, since he would position the turret high enough out of the water so that the centers of the gunports would be over ten feet from the water line. Of course, Eads added, such a remarkable ship would take a long time to build—three years, in fact—and naturally it would be expensive, especially because of the gigantic turret. Still, the government could have this remarkable vessel, a truly "Great Ship," for $6,948,000. Disenchanted, Fox turned to other matters.[19]

Ironically, after the department had uniformly failed in its attempts to build its own seagoing ironclad, a private contractor, William S. Webb, built such a ship from his own plans. Appropriately named *Dunderberg*, this "thundering mountain" of iron displaced 8,000 tons, bristled with four 15-inch and twelve 11-inch guns, and carried 4½-inch armor plate on her wooden sides. Somewhat smaller than the department model, this ironclad ram vessel still measured 368 feet by 75½ feet, and drew 23 feet. Two 100-inch by 45-inch back-action engines powered the single screw which drove the vessel at 15 knots through the open sea.

Originally designed by Webb as a double-turreted ram, the ship soon received a drastic alteration. Lenthall and Isherwood, still convinced that the casemate was superior at sea, firmly opposed the use of turrets and finally convinced Webb of their views. When Fox learned that the turrets would be replaced by a casemate, he strongly objected, assuring Webb that their weight would not be a problem; and in addition, private contractors wanted the work. Despite his protests, Isherwood and Lenthall prevailed. The *Dunderberg* was launched on July 22, 1865, and completed in September, 1866, featuring a casemate 156 feet long, its sides sloping in at a 33-degree angle to deflect enemy fire.

Once the Navy had its seagoing ironclad, it promptly decided there was no longer a need for such a ship. After a period of confused and occasionally bitter discussion, the Navy released the *Dunderberg* to Webb who then sold the vessel, in 1867, to France in whose Navy she served ably for many years as the *Rochambeau*.

Why were the large ironclads proposed by Lenthall and Isherwood never built? "Because," explained Senator James W. Grimes, "the Senate and House of Representatives very wisely refused to do it." The initial encouragement for the seagoing ironclads had come from the commercial cities on the Atlantic Coast who feared the attack of European warships. Congress, however, was unimpressed with the Isherwood-Lenthall school of naval strategy, and steadfastly refused to vote appropriations for such ships. They preferred to rely on the system "now proved to be the best . . . in the world, with the best ships for the purpose"—the monitors. Everybody admitted monitors were admirable for defense, that for harbor protection nothing could exceed them, so why consider any other type of vessel? After all, Grimes inquired, what is the real purpose of our Navy? Is it not to provide us with a defense? "All we want is the monitors to protect our harbors, and then fast vessels to destroy the commerce of a hostile power."

It would be utter folly, the Senator realized, to try to build a Navy which would compete with those of France and England in immense naval battles, so there was no reason for these large, seagoing ironclads. "Our true policy is to protect ourselves at home, and then [with fast, light cruisers] to sweep the commerce of our enemy from the sea," proclaimed the man without whose support the Navy Department could never hope to realize the ambitious program of its leading engineer and constructor.[20]

Congress was not the only obstruction to the seagoing ironclad scheme. Welles had already realized that these ships were not designed to be used against the Confederacy, but against France and England; and when the ultimate military success of the Union became more apparent in 1863 and 1864, the belligerency of these foreign powers noticeably diminished. In all probability, he reasoned, any ironclad actually built would not be completed until after the Civil War was over, and then would there be any real need for it? Welles thought not.

Welles's growing reluctance, Congress' refusal to appropriate money, and Fox's passion for monitors resulted in more than the failure of the seagoing ironclad cruiser program of Isherwood and Lenthall. America was now committed to a naval policy based on passive harbor defense and commerce destruction. Not until the 1890's would the Navy be able to shake itself loose from outmoded concepts which were products of a false tradition of naval success and of a deep desire to avoid full-scale naval competition with the powers of Europe. The beginning of America's naval "dark ages" lay in a refusal to fill the role demanded of a naval power, and as the American Navy clung to the monitor and the unprotected cruiser, its strength ebbed and it slipped irrevocably into a passive stupor, not to revive for a generation.

While Isherwood and Lenthall labored incessantly to build and maintain a reliable steam Navy, it was in the one area where their expert knowledge was shunned that there occurred the episode which Welles, in deep chagrin, branded in his diary as "the most unfortunate transaction that has taken place during my administration of the Navy Department."[21]

During 1862, as the Union armies advanced into the interior of the Confederacy, the Navy provided support by occupying and patrolling the numerous western rivers and by securing lines of communication along the Atlantic and Gulf coasts. It quickly became apparent that the small wooden vessels the Navy had to use were inadequate, because of their lack of protection from land-based rebel artillery. The answer to this problem was the construction of light-draft ironclads which would be invulnerable to enemy fire. As urgent requests poured into the Navy Department for such vessels, the Secretary, acting through his Assistant, Gustavus Fox, turned for help to John Ericsson, the father of the monitor.

On August 4, Fox wrote to Ericsson and stressed the need for an impregnable gunboat for western waters which would draw only six feet. After first doubting whether such a vessel could be built, the famous inventor finally, on October 8, produced a rough sketch for such a vessel which would have a flat-bottomed iron hull, enclosed in a raft of wood; carry three inches of armor on its sides and turret; and receive power from twin engines driving twin screws. His brief description was not sufficiently detailed to be a working plan, but Ericsson was confident that the department could easily build the vessel from what clear and precise data he had supplied.

Normally, such a plan, when submitted to the Navy Department, would go to the heads of the engineering and construction bureaus. As the proposed vessel was to be a monitor, however, the Secretary used a different procedure.

The action of the *Monitor* at Hampton Roads had produced such intense enthusiasm in the North that a "monitor craze" swept the Union, penetrating into the Navy Department itself. Fox became an ardent supporter of Ericsson; and even Gideon Welles, finding in the monitor a salvation from attacks on his administration of the Navy, sought to capitalize on its political utility. There was no room for those who opposed or even questioned the Ericsson ironclad. Isherwood and Lenthall, who many believed had resisted the building of the original monitor, had argued for casemated ironclads, and when using turrets, had preferred the English design. They had established a reputation for opposing monitors, for, as Assistant Secretary Fox recalled, "Lenthall said the monitors would go to the bottom, and Isherwood had no confidence. . . ."[22] The department, when considering the Ericsson plan for light-draft monitors, naturally hesitated to allow these bureau chiefs to supervise the construction.

The Navy Department had established, in July, 1861, an office in

New York city to supervise the construction of all ship and engine work done by private contract for the Navy along the Atlantic Coast. When its functions expanded to include the building of ironclads as well, this office became known as the "monitor bureau" and enjoyed only a nominal connection with the bureaus of construction and engineering in Washington. As it soon appeared, this independence from bureau control created a serious division of function and a breakdown of communication between responsible individuals. It was in these circumstances that Welles turned Ericsson's light-draft-monitor plan over to the New York office.

In charge of the "monitor bureau" as general superintendent of ironclads was the aged, infirm Rear Admiral Francis H. Gregory. Seventy-three years of age in 1861, this old line officer, who had served in the Navy since 1809, had been brought out of retirement, given increased rank, and set at a desk in the New York office to administer the steamer and ironclad building program. As his general inspector of ironclads (also known as the general inspector of steam machinery for the Navy), the department assigned Chief Engineer Alban C. Stimers, a vain, clever, and immensely ambitious naval engineer, who had distinguished himself by supervising the construction of the *Monitor* and then serving on that vessel throughout the battle of Hampton Roads. A close associate of Ericsson, Stimers was a specialist in ironclads and the technical expert for the "monitor bureau."

Understandably annoyed at this intrusion into their professional domain, Isherwood and Lenthall stood helpless, recognizing, as one leading shipbuilder of that era, Charles H. Cramp, of Philadelphia, recalled, that they "had no power to antagonize the monitor craze successfully." Cramp stated further that the bureau chiefs had been "entirely set aside, and practically disappeared from the scene as far as new constructions were concerned."[23] Stimers, who had previously complained if Isherwood did not write him regularly, now became aloof and even truculent towards his bureau Chief. Pathologically sensitive, he became convinced that both Isherwood and Lenthall were blinded by "personal enmity" toward him, and would go to any length simply for the pleasure of defeating and frustrating him because his new position challenged their authority.

Supporting Stimers in his rages against the bureau was the crusty Swedish inventor who hotly resented any interference or modification in his own work. By the end of 1862, Ericsson could write to Fox,

> . . . that unless the malign influence of the Engineer-in-Chief can be wholly removed from the Monitor fleet, to which he is bitterly and openly hostile, the Nation will have much cause for complaint.[24]

Worried by Ericsson's belligerence and wishing to avoid a serious dispute, Fox hastily wrote the inventor, asking him to control Stimers' outbursts, as well as his own. There was no reason for such attacks on Isherwood, who certainly had no "malign influence" against the monitor fleet, as Ericsson maintained. Stimers was endangering his naval career by openly attacking his bureau Chief, and Isherwood was far too busy with his own work to have to submit to such abuse from either Stimers or Ericsson. With this admonition, the inventor's irritation apparently subsided, and for a time he buried his resentment in the intensity of his labors.

Already overburdened with the designing of several classes of monitors, Ericsson had no time to work out in detail the plans for the light-draft monitors. The department, therefore, turned the entire job over to Stimers, and directed him to establish an office adjacent to Ericsson's so that he could consult the inventor on all the necessary changes and modifications to the original rough plans.

Seizing this opportunity, Stimers soon had a combined engineering and construction bureau of his own in operation, staffed with engineers, draftsmen, and clerks whose number almost equalled in size the entire office force of all the Navy bureaus combined. Intoxicated by the importance of his work and impatient with red tape and supervision, Stimers largely ignored the bureaus in Washington, paid mere lip service to his administrative superior, Admiral Gregory, and freely went his own way in designing the monitors.

By February, 1863, Stimers had prepared detailed plans and specifications, but when it came time for the Navy Department to advertise for bids, he decided to retain control over his monitors. Depositing, as Lenthall wrote to Welles, "nothing more than outlines" with the Bureau of Construction and Repair, he forced the Navy Department to include only a brief, general description of the vessels in its advertisement of February 10. Interested builders, therefore, had to come to Stimers to obtain precise information on the vessels before they could bid.[25]

Before opening the bids, Fox nervously wrote to Ericsson, February 21, 1863, asking for his assurance that the plans were correct. He was staking the reputation of the Navy on these vessels, Fox said, and he had to make certain that everything would go as planned. Stimers may have been supremely confident in his abilities, but John Ericsson felt differently. Not having paid attention to Stimers' work, he suddenly awoke to the fact that something was very wrong. After examining Stimers' plans, Ericsson hastily telegraphed Welles, February 24, warning the Secretary that his "leading principle" had been "frittered away by changes," and he could not be held responsible for the final product. Writing Fox on the same day, the proud and stubborn inventor was deeply mortified, admitting, "I cannot find words to express my embarrassment." One look at Stimers' work and his own course was clear, said Ericsson: "that of repudiating all responsibility. . . ."[26]

The Navy Department did not hesitate for long, despite Ericsson's warnings. All they had received from the inventor was a rough sketch, whereas from Stimers they had an elaborate and full plan of the vessel from which builders could apparently construct the vessel with ease. On the basis of its completeness, Isherwood and Lenthall stated that they preferred Stimers' plan, although they did not inspect either his drawings or his calculations. There had been a number of changes to Ericsson's original plan, such as the inclusion of water tanks which would afford the vessel greater buoyancy when pumped dry, a suggestion of Admiral Joseph Smith, chief of the Bureau of Yards and Docks. A casual examination by department officials resulted in the approval of Stimers' plans which incorporated these new features.

At one point, Welles finally turned to his engineering expert, asking Isherwood's opinion of the monitors' machinery. However, Isherwood had no drawings and only a handful of statistics to work from; and as he understood it, he had been requested only to determine the probable speed of the vessel, rather than to evaluate the machinery design. On this basis, Isherwood compared Ericsson's original machinery with that proposed by Stimers, and told Welles, in a letter dated April 7, that the modified plan would provide 21 per cent more power, based on a change in the size of the engine and the boiler. Apparently encouraged by Isherwood's report,

Welles allowed the contracts to be made, and between March and May, 1863, twenty light-draft monitors were put under construction in ten different cities.

During the spring and summer of 1863, Stimers was unable to give his close attention to the construction of the monitors. Not only were there a number of other iron vessels being constructed under his supervision, but he had also become embroiled in a dispute with the commander of the South Atlantic Squadron whose ironclads were attempting to force the harbor at Charleston, South Carolina. Rear Admiral Samuel F. du Pont, stung by Stimers' public criticisms of his use of monitors against the Confederate batteries, preferred charges against the engineer and demanded a court-martial. Stimers was thus unable to concentrate on the light-draft monitors, and left most of the detailed work to the builders.

Had Ericsson also been supervising these vessels while they were being built, there might not have been any problem. Unfortunately, Ericsson had refused to assist with the monitors. After his initial repudiation of Stimers' modifications, Ericsson had then argued with the engineer over a trifling technical matter. Their difference of opinion was serious enough to create an open breach, and Ericsson thereafter refused to speak to his former associate. The entire planning of the monitors was now in Stimers' grasp, and there was no one to check his work. Ericsson, who would not correct Stimers' mistakes, was concerned only for his own reputation. Fearing that he would be blamed if the monitors proved worthless, he continued his warnings to the department and disclaimed all responsibility.

Stimers now gloried in his freedom from the supervision and interference of Ericsson, Isherwood, or Lenthall. Rear Admiral Gregory, tired and unable to handle the work of his office, allowed his impetuous associate to take over to the point where the engineer treated his superior as little more than a front man. Mildly resentful, but unwilling to make an issue of their relationship, the old officer was content to sit back and observe the maneuverings of his colleague. Still perceptive, despite his age and illness, Gregory observed the engineer's passion for monopolizing the ironclad program and warned Fox, in a letter dated June 27, 1863, that "Stimers affection for the Monitors is so intense that he cannot bear the least interference with whatever concerns them. . . ."[27] This fact had become quite apparent when the department allowed Chief Engineer James W. King to supervise the construction of gunboats built at St. Louis. Stimers became furious at this encroachment, the Admiral explained to Fox, because "a joint of his tail had been cut off—occasioning considerable agony."[28]

If mistakes appeared in his plans, Stimers brushed them off as inconsequential. In August, 1863, he wrote to Fox asking for authorization for a number of necessary changes in the monitors' design, and added he hoped that the department would forget any previous design problems. "If I had been timid about taking responsibility," Stimers boasted, "we never would have got them started."[29]

Responsibility—and the high offices which went with it—were ever before Stimers' eager gaze. When Du Pont badgered him about the ironclads at Charleston, Stimers decided that the best way to humble the Admiral would be for the Navy Department to show its confidence in him by awarding him the commission as engineer in chief and then widely publicizing this act. That he would also rid himself of a professional rival did not escape Stimers. The more he supervised the ironclad program, the more he became convinced that the wrong man was head of the steam bureau in Washington. By early 1864, he arranged for the distribution of a scurrilous pamphlet, written by E. N. Dickerson and entitled *The Navy of the United States,* attacking Isherwood and the administration of his bureau; and boasted to Fox, in a letter dated January 6, 1864, that it would be placed on the desk of every congressman.

When it appeared that he could not dislodge Isherwood from his office, Stimers produced a new plan. In February, 1864, he proposed the establishment of a new "Bureau of Ironclad Steamers" to rank with the existing bureaus of construction and engineering. Lenthall was not really incompetent, Stimers allowed, but the naval constructor had proved unequal to the task of producing an ironclad, probably because he had always opposed such a vessel. Later, when the demand for ironclads became great, Stimers continued, Lenthall and Isherwood had frittered away the time with their "grand designs." It was now time that a "practical and scientific engineer especially skilled in the construction of ironclad steamers" took over, and Stimers knew just the engineer for the job.[30]

The Navy Department did not have to create a new bureau to find out what would happen when Stimers took over. In May, 1864, the *Chimo,* first of the light-draft monitors to be launched, displayed the results of his engineering and scientific skill. The vessel could barely float. Before her turret was installed, the *Chimo*'s decks were only three inches above the water line, a foot less than planned. Without her coal and ammunition her decks were awash, and with a full load, her hull ran slightly submerged and at a speed of only three knots.

Perturbed at this discovery, Fox on June 3, 1864, wrote to Ericsson for help, remarking with commendable understatement, "Our friend Stimers falls short somewhat in his calculations on the light drafts."[31] Ericsson, still refusing to deal with Stimers, would have nothing to do with these vessels, which, he now regretted, bore the name of monitor. Stimers, learning of Ericsson's refusal to help, promptly accused the great inventor of doing "everything in his power to crush what he chooses to consider a formidable rival." Stung by Ericsson's treatment of him as a "charlatan engineer," Stimers warned Fox, in a letter dated June 8, 1864, that the inventor now wanted the light-draft monitors to fail, since

If he can now prevent these vessels becoming in any way servicable Fox is disgraced and Stimers utterly annihilated; the fame of Ericsson shining with renewed splendor because of the failure of those who had the audacity to suppose they could improve upon his plans.[32]

In addition, Stimers now conveniently discovered that the failure of the vessels was not actually his fault at all, but rather that of one of his young engineers who had made a serious miscalculation which Stimers, busy with his other duties, did not notice.

Secretary Welles was not deceived by these various attempts to distract attention from the fact that the ships which had failed were Stimers' responsibility. Although Stimers and Fox urged him not to worry, Welles had received too many reports from other officers to have any illusions about the matter. Up to this time, Welles had assumed that Stimers was working closely with Ericsson. Now, realizing the true state of that relationship, the Secretary remarked, in a letter to Fox dated June 10, 1864, that Stimers "must not be tenacious in holding on to admitted errors"; and to make certain that this situation did not occur, he removed Stimers from his duties as general inspector of ironclads.[33]

His pride wounded, Stimers went to Washington to look for solace from Fox. Refusing to admit that his removal might be the result of his own errors, he darkly accused Ericsson and the bureau chiefs of persecution. Fox was his one friend in the department, he insisted, and only because of this friendship had he come to offer his services to Fox "in weathering the storm which it would appear I have brought upon the Department."[34]

Stimers' habit of covering up his mistakes by accusing others of evil motives greatly exasperated Ericsson. The inventor grumbled to Fox, July 23, 1864, that Stimers "is disgusting all loyal people by his levity in the light draft monitor disaster. He appears to exult in the matter, and repeats the falsehood everywhere that he has nothing to do with it, although ... [nearly] ... every part to the minutest detail is his own planning. ..."[35] Fox agreed that Stimers had committed "gross blunders," but he hoped that this would not prevent Ericsson, whose "ingenuity and genius" had created the monitors, from rectifying the situation.[36]

The Secretary of the Navy, wise in the ways of men and of the workings of his own department, saw that the blame for this affair rested not only on Stimers. He had made a mistake in trusting Stimers, Welles admitted, but this was because he had been "deceived" into assuming that Ericsson and Lenthall were properly supervising the work. Once realizing his error, Welles immediately called in the Chief of the Bureau of Construction and Repair, who blandly informed the Secretary that he was not responsible for the monitors and, furthermore, knew nothing about them. Furious at Lenthall's lack of concern, Welles told him that this explanation was not acceptable; that it was Lenthall's duty to know what was going on in his special field, and he could not plead ignorance as an excuse.[37]

Although startled by the Secretary's anger, Lenthall was not at all displeased at the outcome of the light-draft-monitor affair. Well aware that Fox and Stimers were responsible for the result, Lenthall felt that this was a just reward for their temerity in refusing to consult experts such as himself, and for their selfish desire to claim for themselves alone the fame which would follow from the anticipated success of the vessels.

Welles quickly realized that Isherwood and Lenthall, mortified by the creation of the New York "monitor bureau," had decided to give Stimers enough rope to hang himself. As Welles discovered, the two bureau chiefs thus "culpably withheld ... information of what was being done," hoping that Fox and Stimers, in their enthusiasm, would go too far. When the light-draft monitors did prove to be a failure, Welles called in Fox and Stimers for an explanation. Suddenly frightened and unwilling to admit their own rashness and incapacity, they placed the blame on Isherwood and Lenthall. The bureau chiefs, now accused of responsibility for a faulty design, when in fact they had deliberately ignored the monitor plans, decided to admit their deliberate lack of participation and disclose the entire truth. "The whole thing," Welles sadly concluded, "was disgraceful."[38]

Unfortunately, news of the failure of such a publicized class of vessels could not be confined to the Navy Department. Enemies of the administration and of the department seized upon the episode and demanded a full-scale congressional investigation. On June 29, 1864, a resolution passed Congress ordering the Committee on the Conduct of the War to include the light-draft-monitor affair among its various investigations. This body of congressmen, basically hostile to Lincoln's wartime administration, was stigmatized as a group of "narrow and prejudiced partisans, mischievous busybodies" whose "mean and contemptible partisanship colors all their acts" by Gideon Welles, who was not exactly impartial himself when under attack.[39]

The congressional investigation of the light-draft-monitor affair began in late December, 1864. All the leading participants gave testimony, including Isherwood and Lenthall, who both retestified when confronted with the account given by Stimers. He claimed that Isherwood had wished to change completely Ericsson's machineery plans, and had taken an active part in redesigning the engines and boilers of the light-drafts. Therefore, the bureau chief shared in the responsibility for the failure of the vessel.

Nevertheless, declared Stimers, Isherwood and Lenthall were both "inimical to these vessels." Isherwood had brusquely refused to give him more naval engineers to inspect the light-drafts because their construction was not being done under the auspices of the Bureau of Steam Engineering. He tried to appeal to their "old friendships and old association," but Isherwood rudely turned him down, even when the engineer implied he would soon be in a position to do his Chief a few favors. Stimers thus concluded that he could not hope for help from Isherwood or his colleague Lenthall. "I always felt that it was a regular fight," he recalled, "that we had to conquer them before we could get them to do anything." In his opinion, the success of the Navy's ironclad program depended upon his ability to beat down the bureaus on the one hand and the contractors on the other.[40]

Isherwood's interpretation of their relationship was not the same. He testified that Stimers, far from appealing to his friendship for assistance, had sauntered into his office and simply showed him what was being done on the monitors, with no attempt to ask Isherwood's permission or approval. Consequently, aside from the estimate of the monitors' speed, Isherwood insisted that he had nothing to do with the program, and righteously defended his official ignorance of Stimers' actions. Stimers' manner toward his Chief, far from being deferential, had been of "a superior condescending to explain his plans to an inferior, in order that he might receive his admiration." That he had appealed to his Chief for help as a personal favor was simply false, Isherwood declared, especially since "he was endeavoring, by the basest arts, the vilest calumnies, and the most dishonest practices, to supplant me as Chief of the Bureau of Steam Engineering...."[41]

The part of Stimers' testimony and general behavior which most annoyed Isherwood was the engineer's "shocking moral obliquity" in attempting to shift the blame for the monitors to others. When the failure of the light-drafts became apparent, Stimers had the "matchless effrontery to attempt throwing the responsibility of the parentage of his wretched abortions upon the two mechanical bureaus of the navy."[42] To the proud and dedicated Engineer in Chief, this was the most reprehensible part of Stimers' behavior:

That a person should not have ability equal to the performance of a task which his self-conceit makes him undertake, is not uncommon; but it is very uncommon to find so little manhood as not only to shrink from the responsibility of the failure when it comes, but the baseness to attempt screening himself by falsely charging it upon the well-won reputation of others.[43]

The Committee on the Conduct of the War agreed that Stimers was primarily responsible for the failure of the light-draft monitors, but suggested that the Navy Department was also partly at fault, by failing to supervise adequately the design and construction of the vessels. Welles agreed with this, to the extent that he held Fox, rather than Stimers, as the main offender. As a high-ranking department officer, it was Fox's responsibility that nothing should go wrong, and his continual correspondence with Stimers deprived Fox of the excuse of ignorance. Stimers was implicated in the monitor disaster, but only as a subordinate who had become "intoxicated, overloaded with vanity," and who took on more than he could handle. "More weak than wicked," Stimers should not be judged too harshly, Welles felt, especially as he had served admirably on the *Monitor* and had shown engineering ability, although Isherwood and Lenthall vigorously denied this.[44]

Fox had disappointed him and caused him considerable annoyance, but the Secretary could forgive his impetuous and irresponsible subordinate. Fox's problem, Welles realized, was his eagerness to "get his name in the history of these times" and to appear as an equal with Welles in the Navy Department. The Assistant Secretary wanted to be "all powerful" with naval officers, and often sent out Welles's orders as his own, a practice which greatly irritated the bureau chiefs. The light-draft-monitor affair was an unhappy demonstration not only of Fox's ambition to run the whole show, but also his readiness to "shun a fair and honest responsibility for his own errors" when things went wrong.[45] The sage and venerable Secretary could tolerate such weaknesses, since Fox worked well with him, was essential in his job, and did not oppose or interfere with Welles's policies.

In addition, Fox was immensely valuable as a liaison between the Secretary and his naval officers. During the war, the Assistant Secretary established a remarkable correspondence with virtually every important officer in the Navy who was away from Washington. His uniformly receptive and sympathetic manner brought him a flood of "private and confidential" letters from men who were his friends or from those who thought they saw in the Assistant Secretary a means of achieving their own ends. Gideon Welles, old, stern, and forbidding with his flowing white beard and austere manner, seemed unapproachable and unsympathetic to many officers who thus turned to the younger, more energetic, and very congenial "Gus" Fox. Moreover, Fox was as popular outside the service as he was within it. His correspondence with Ericsson led to a firm and lasting friendship which won Ericsson a powerful advocate for his monitor theories and a valuable friend at court.

At times, Fox's sense of self-importance, tremendously bolstered by the flood of confidences he continually received, betrayed him. In the light-draft-monitor episode, he was perhaps deceived by the buoyant optimism of Stimers; but, if so, it was because he so deeply wished to believe that the monitors would prove an enormous success. His star would rise with that of the monitors, Fox believed; Stimers was the last person to disillusion him.

But even Stimers finally ran out of confidence in his own work. When the light-draft-monitor *Tunxis* was finally put in shape to be commissioned in 1864, he was ordered to this vessel as chief engineer. When boarding the *Tunxis* he discovered a plaque set into the vessel which stated that the builders had constructed the vessel "from designs prepared by Alban C. Stimers, Chief Engineer of the United States Navy." Reflecting on the reputation of this class of vessel, Stimers for once became modest of publicity and proceeded to cut his name out of the plate with a cold chisel. As John Lenthall smugly summarized it,

So far from submitting to be instructed by Mr. Ericsson, he assumed to be his rival, and in the endeavor to imitate him underwent the fate of the frog who attempted to expand himself to the bulk of the ox.[46]

Isherwood's role and behavior in this episode are worthy of note. He and Lenthall had been rudely ignored and their professional qualifications slighted by the creation of the New York office and its monopolization by Stimers. Without Isherwood's permission or knowledge, a number of engineers were assigned directly to Stimers' office, no attempt having been made to clear this action through the Bureau of Steam Engineering. During the design of the light-draft monitors, Isherwood was rarely consulted, and when he was it was only in an offhand way by a subordinate who obviously desired Isherwood's own job.

Unaccustomed to such treatment, Isherwood joined Lenthall in a hands-off policy designed to teach a lesson to those in the department who undervalued the knowledge and experience of the two bureau chiefs. Losing sight of the Union's need for light-draft monitors, Isherwood and Lenthall selfishly gratified their wounded sensibilities and let the vessel program fail. A proud and stubborn man, Isherwood for once let his personal feelings overcome his sense of duty, with results which hurt the Union war effort, embarrassed the Navy Department, and did no credit to himself.

## IV. The Unwelcome Pioneer

Benjamin Isherwood's naval career spanned a period of forty years, but for nearly twice that time he was a practicing engineer. His scientific work began in an era when *engineer* meant engine-driver, and did not end until engineering was a recognized profession of highly educated men who dealt with the myriad complexities of applied science. During this period he was one of engineering's true pioneers; a man who shattered traditions and challenged dogmas, while both articulating and demonstrating in his own work the invaluable attributes of the scientific method as applied to mechanics. Isherwood bridged the chasm between theoretical postulates and practical utility, and in so doing, was a major influence in raising his profession from the humble occupation of shopworkers to the specialized field of highly trained experts.

Just as the naval engineers of the mid-nineteenth century were engineers far more than naval officers, Isherwood's naval career also revolved around his engineering work. He joined the Navy not from a love of the sea or ships, but because it offered a fine opportunity to a young engineer. His years of dedicated service proved to be of incalculable value to the Navy, but not for his exploits at sea or for producing heroic expostulations which would grip the hearts of future generations of Navy men. His entire colorful naval career never disguised the fact that he was first of all an engineer.

Steam engineering in the mid-nineteenth century was still in its infancy. Despite the work of James Watt over half a century before, the structure and operating efficiency of the steam engine had improved only to a slight degree. The reason for this lack of development seems to have been caused by the inability of physical scientists to translate their theories into practical application. The modern kinetic theory of heat, attributing it to the agitation of particles of matter, did not appear until Joule's experiments in Manchester, England, between 1840 and 1850. Even then, the results of his work were slow to spread, so that well into the 1850's authorities still clung to the old caloric theory which hypothesized the existence of a tenuous fluid which entered and left bodies without changing their weight.[1]

The first systematic account of steam-engine theory did not arrive until 1858, when the Scottish physicist W. J. Macquorn Rankine produced *The Steam Engine and Other Prime Movers*. Although this book has been hailed to this day as a landmark in the history of steam power, Rankine divorced the theory of the steam engine from its operation, making assumptions which were entirely untenable in practice, and which rendered his work quite impractical for the engineer who sought an applied theory of the heat engine. Nevertheless, this and similar works would be used into the 1870's as textbooks for engineers.

While scientists failed to deal with the practical application of theories, engineers generally failed to understand the principles underlying the operation of their engines. Steam engineering was a "rule of thumb" profession, the work of unlettered, unscientific men who were no more than glorified mechanics. There was little written on steam engineering which could help them, because those men who did write based their engineering precepts on purely theoretical knowledge, while those who worked with and experimented on engines were largely self-taught and displayed little ability or enthusiasm for writing, as Isherwood pointed out in *Experimental Researches in Steam Engineering*. Engineering was primitive because there were few recorded facts on engine operations. True engineering knowledge and skill could come only when sufficient experimentation finally amassed the indispensable data for sound conclusions.

As a result of the failure to develop engineering knowledge and sound theory, steam engineering, in 1860, was, in Isherwood's words, in a "deranged state."[2] With little improvement over the design of Watt, a good, standard, marine engine of the early 1860's ran on only twenty to thirty pounds of saturated steam from its low-pressure boilers, had little or no insulation for its cylinder, and performed with discouraging inefficiency. Unable to apply the theoretical works on thermodynamics to their engines, the steam engineers of that day usually contented themselves with minor refinements of their primitive machinery. Those more adventurous souls who hopefully seized on theoretical postulates as a panacea produced monstrous abortions of engines, through their clumsy efforts to translate imperfectly learned or fallacious theory into sound practice. Ironically, even when the theory was basically sound, it often proved harmful to steam engineering because of the inept mechanical structures so trustingly contrived on its principles.

Isherwood learned wholly from experience. Since technical or engineering schools were not available to him, a solid grasp of basic mathematics was his only intellectual preparation. To be an engineer, he had to go directly to the engine. Laboring in the machine shops of the Utica and Schenectady Railroad and of the Novelty Iron Works exposed him to the details of the reciprocating steam engine. By the time he entered the Navy in 1844, he was thoroughly familiar with the workings of such mechanical devices, and his naval apprenticeship would further his study of the marine steam engine until he had mastered the intricacies of propulsion from boiler to propeller.

He soon proved to be more than just another young steam engineer. At a point early in his life he had developed a passion for detail. His earliest written work was gorged with data, a characteristic which would only intensify as Isherwood pursued his career.

Furthermore, he soon developed an interest in engineering which went beyond the field of basic mechanics. When the material in available technical publications proved inadequate, he turned to the writings of the acknowledged masters in physics and engineering in order to find solutions to basic problems he had encountered in engine design and operation. The area which particularly intrigued him was the new field of thermodynamics, and Isherwood avidly studied the pioneering works of Tyndall, Joule, Mayer, Rankine, Clausius, and Hirn. As the writings of several of these scholars had been published only in French, Isherwood applied his skill in this language to good use, translating many articles which had never appeared in English-language journals.

Isherwood began early to publish the results of his experiments. During the 1850's a number of short articles and notes appeared under his name in the pages of the *Journal of the Franklin Institute,* a thoroughly professional publication whose standards for scholarly, scientific reporting were then, as they are today, among the highest in the world. Isherwood was to write for this journal over the next half century, contributing well over one hundred articles on every aspect of mechanical and naval engineering. His early contributions were mostly on the specifications of steamers, and it was his experiments and collection of data in this field which led him to write his first book.

*Engineering Precedents for Steam Machinery,* the first volume of which appeared in 1859, contained a series of technical essays on British gunboats and American screw steamers, and included a comparison between American screw and paddle-wheel vessels. In his off-duty hours in Hong Kong, China, and at Simon's Bay, South Africa, during 1857 and 1858, Isherwood had busied himself examining several British vessels, checking their logbooks and indicator diagrams and interviewing their engineers.[3] In his typical fashion, Isherwood accumulated a formidable amount of data which he later patiently collated and set down in tables which ran on for pages in his book. *Engineering Precedents,* however, was not just a dispassionate collection of data. In dealing with steamers of the American Navy, Isherwood took this opportunity to attack the designs, calculations, and conclusions of C. H. Haswell, former engineer in chief of the Navy.

Isherwood's first major publishing effort had a favorable reception. Encouraged, he quickly assembled further data and brought out volume two of *Engineering Precedents* several months later. Here he reported on Navy Department experiments made on various types of coal, and then warmly endorsed the vertical water-tube boiler designed by Chief Engineer Daniel Martin, on the basis of experiments made to compare this boiler with the horizontal fire-tube boilers generally used in steamers.

The coal tests, he believed, were "the only rigorously comparative and reliable experiments ever made for this purpose."[4] In contrast to the normal laboratory trials, the testing of the coal had been made under actual working conditions so that the results would have practical value by providing usable data for the determination of coal consumption and engine efficiency.

The boiler comparison, a *"pièce de résistance* in marine boilers,"[5] as Isherwood modestly described it, demonstrated the far greater efficiency of Martin's boiler. So convinced was Isherwood of its superiority that he doggedly supported this design and ordered these boilers installed in American warships during the Civil War, despite continual criticism and even accusations that he was receiving from

Martin a percentage of the royalty payments. The Martin boiler, Isherwood later explained, was opposed in Great Britain and America, mainly because of the influence of large steam-engine builders. These manufacturers were indifferent to the engineering merits of Martin's design because, in constructing boilers, they disregarded the efficiency of their product. The Martin water-tube boiler was much heavier, bulkier, and more expensive to construct than any fire-tube boiler which produced equal amounts of steam. Since builders received a fixed sum to produce either a given horsepower or a specified speed in a vessel, regardless of the coal consumed in producing such a result, they naturally wished to build the cheapest type of boiler. Isherwood, as chief of the steam bureau, was able to specify the boilers to be used in government vessels, and he brought the wrath of builders on his head by demanding the boiler which would give the most efficient operation.

The most significant of the experiments published in *Engineering Precedents,* Volume II, started Isherwood off on a thorny path from which he never wavered, despite the acrimony which would assail him for the next decade. With his deep suspicion of engineering practices based on theoretical postulates, he determined to question the theory of steam expansion, a concept which had so captivated the imaginations of steam engineers of the 1850's that it had become elevated into virtually unassailable law.

The purpose of expanding steam within a cylinder was to utilize its elastic properties, when in a highly compressed state, to do much of the work in driving the piston of a reciprocating engine. Without employing expansion, steam at a normal pressure of twenty-five pounds would have to be injected into the cylinder throughout the entire length of the piston stroke, and then be entirely exhausted while the piston, impelled by steam injected into the opposite end of the cylinder, made its return stroke. As far back as Watt, engineers had seen the possibility of economizing on the amount of steam used by cutting off its injection before the piston stroke was completed, thus allowing the compressed vapor already in the cylinder to continue impelling the piston as this steam expanded.

The logical extension of this argument was that very little steam might be injected into the cylinder, and the piston would thus be driven almost entirely by the expanding force of the vapor. Much less steam would be required, and fuel costs in its generation would consequently be reduced. However, as it expanded in volume within the cylinder, steam would lose pressure proportionately. Therefore, its initial injection would have to be at high pressures in order to assure a sufficient average force against the piston. Nevertheless, engineers generally believed that the "use of steam expansively" would be carried out to an extreme degree, thus promising enormous economy in engine operation.

Benjamin Isherwood, however, had his doubts. He recognized the merits of using the expansive properties of steam, but he questioned the real economy resulting from such use. Theoretically, steam, as a gas, decreased in pressure in exact proportion to its increase in volume, assuming its temperature remained constant. This hypothesis, dogmatized in the early nineteenth century as "Mariotte's law," encouraged the extreme use of expanded steam, since all that engineers had to do was to start with a high initial steam pressure, cut off its admission at a small fraction of the completed piston stroke, and let the expanding vapor do the rest of the work. Isherwood was skeptical of this theory because it did not take into account the actual operating conditions of the engine. It considered only three elements: pressure, volume, and a constant temperature; and any change in one variable, according to Mariotte's law, would produce a mathematically precise reaction in the other.[6] This theory might suffice for a laboratory experiment, but for an actual steam engine under working conditions, there were other factors which could modify the law of expanding steam.

With this thought in mind, Isherwood decided to use an ordinary steam engine and boiler to determine the power produced by cutting off the steam at one-fifth of the piston stroke compared to that when no cutoff was used until the end of the stroke. According to Mariotte's law, the same amount of fuel would produce 1.515 greater horsepower with the one-fifth cutoff, by allowing the steam to expand to five times its original volume. The actual results he obtained greatly satisfied the skeptical engineer. He discovered that only 16.8 per cent more power would be produced by this large degree of expansion than if there had been none at all. This "enormous" discrepancy between the "theoretical prediction and the practical result" convinced Isherwood that he was on the right track, and that scientists and most engineers, as he had expected, were following a false trail.[7]

The discrepancy he had observed, Isherwood explained, resulted from a number of related factors. Of the generated steam injected into the cylinder, not all of it actually worked against the piston itself. Some of it would be lost by condensation from heat loss within the cylinder. A small portion would escape through leakage past the piston and the valves, and some would flow into the spaces provided by the steam ports. Also, the piston would not be entirely unresisting to steam pressure, since there would be a certain amount of "back pressure" caused by uncondensed and unexhausted vapor remaining from the previous injection in the other end of the cylinder.[8]

Of these results, the most significant for Isherwood was the effect of steam condensation. As its pressure decreased and volume increased, the temperature of the steam did not remain constant, but tended to fall. Ultimately, a sufficient drop in temperature would cause the steam to condense into water, and this condensation would, in turn, decrease the pressure of the remaining steam against the piston, resulting in a direct power loss. Those who extolled the steam expansion theory admitted the loss of heat but maintained that temperatures would never drop low enough for condensation to occur. They failed to take into account a very important practical factor, Isherwood noted. The metal walls of the cylinder were not perfect insulators, and thus drew heat from the steam and radiated it out to their cooler exterior surfaces. The cylinder might be insulated, of course; but in the late 1850's and early 1860's, such insulating by steam jacketing or similar means was inefficient and expensive, and most of the marine engines did not have this feature. In addition to the cylinder itself taking heat from the steam, there was a considerable heat loss even before the steam reached the cylinder. The boiler, connecting pipes, and the steam-valve chest all offered additional means of losing heat through conduction and radiation.[9]

In addition to the loss of heat in the cylinder through conduction and radiation, there was a loss through the process of expansion itself. As the steam rapidly expanded, particles of vapor near the cooler cylinder walls would lose so much heat that they would begin to condense. Their proximity to the remaining steam, however, would bring about their vaporization back into steam, but at a price. To turn these condensed particles back into steam required heat, so the process of vaporization would lower the average temperature of the whole. This fall in temperature, in turn, would allow other steam particles near the cooler metal surfaces to condense; and the cycle would begin again, always taking more heat from the mass of vapor. Its loss of heat, together with its increasing volume, produced a continuous decline in the pressure of the steam, and here lay the danger of excessive reliance on the principle of expansion. If the expanding steam lost heat too quickly, the average pressure against the piston would fall too low to provide adequate power. For the saving in fuel costs, the engineer would pay an exorbitant price in the loss of power, and the efficiency he had so greatly prized would vanish.

On the basis of these results, Isherwood insisted that steam was not a gas, but a highly unstable vapor which could not be governed by a law applying to a gas. For this reason, he determined to attack the hypothetical "law of MARIOTTE, which though rigorously true for the imaginary conditions upon which . . . based, are yet subject to such wide modifications by the practical conditions actually encountered with steam machinery, that the predicted advantages may be lessened to any extent, and may even disappear under very unfavorable circumstances, while under the most favorable ones possible there will always remain an enormous discrepancy."[10]

This time his book received more than casual notice. The *Journal of the Franklin Institute* considered it to be far more valuable than his first volume because of his experiments on working steam expansively. Thorough and painstaking as his methods were, the editors mainly commended Isherwood for being "bold in enunciating his ideas and deductions," especially since his work conflicted with generally accepted theories.[11]

Isherwood was convinced that he had made a major contribution to the field of engineering, because so seldom were there any complete records of reliable experiments. Of all the scientific reports of his day, only a few included full descriptions of the apparatus used, the entire procedure of conducting experiments, and the reasoning on which conclusions were based. Too many experimenters relied on their fame rather than upon the concrete and complete evidence they could present. A reader, he argued, should not have to rely solely on the integrity of an experimenter. Data should be "full to the minutest detail," so that the reader could follow step by step. There was even a valuable by-product of such detail. Often data might incidentally be obtained which had no relevance to the particular experiment in question, but still had intrinsic value. For this reason, Isherwood urged, all data should be included in a report, even when not pertinent.[12]

His work on steam expansion, together with similar analysis on the distribution of engine power, served a far greater purpose than simply modifying an accepted theory, Isherwood believed. He hoped his experiments could restore a measure of sanity to the engineering profession, currently in the grip of a mania for patent cutoffs which promised great economies from a modicum of steam—a preoccupation he dismissed as "the most exaggerated nonsense." What he wanted to do was to simplify the steam engine and make its operation reliable by basing it on practical principles. His was the duty to check the excesses of experimenters who took the bull by the horns and darted off to unrealistic and impractical extremes. There was indeed a difference, Isherwood concluded, between mechanical ingenuity and mechanical judgment, and it was his duty to make this fact clear.[13]

When Isherwood went to Erie, Pennsylvania, in November, 1860, he already had established and published his views on steam expansion. The experiments he and his board of naval engineers were to make on the *Michigan,* however, were to give his views their first wide publicity; and it was the results he obtained here which later became the focus of attacks on Isherwood's position. His orders were to make a set of experiments on the valve gear of the *Michigan*'s engines and to determine the evaporative efficiency of the old paddle wheeler's boilers. The experiments were to determine the relative economy of fuel in relation to the power developed, based on the use of steam at varying measures of expansion.

Isherwood conducted the Erie experiments from December 30 to January 28 of the following year. Using only the port engine of the *Michigan,* he ran the machinery at a constant speed, using careful measures of coal and water. The only variable in his procedure was the place in the piston stroke where the steam would be cut off. Isherwood determined to set his cutoff at $11/12$, $7/10$, $4/9$, $3/10$, $1/4$, $1/6$, and $4/45$ of the piston stroke. At each cutoff position he ran the engine continually for 72 hours, keeping the steam pressure in the boiler between 19.5 and 22 pounds.

After assembling his data and carefully analyzing the results, Isherwood reported that the optimum place for the cutoff for this engine was at 7/10 of the stroke. This meant that the steam could be best utilized when expanding only 43 per cent beyond its original volume, which was "scarcely recognized as working it expansively," Isherwood maintained. The results of this experimentation, he felt, were "conclusive against the popular belief in favor of high measures of expansion," and therefore it would be utter futility to expand steam more than 1½ times its original bulk. Indeed, expanding it 3 times created a loss of economy, and all high rates of expansion showed less economy than no expansion at all. Listing 10 different factors which forced a modification of Mariotte's law when applied to steam, Isherwood said that all but 2 were greatly affected by the degree to which steam was expanded, and they became rapidly more potent causes for power loss as the measure of expansion increased.[14]

This being a practical rather than a laboratory experiment, according to Isherwood's firm beliefs, there were several aspects which drew outraged protests from engineers who disputed the validity of his claims. He insisted on using a typical marine engine of the early 1860's, which meant that the cylinder had no insulation. The fact that steam jacketing was well known and extensively utilized elsewhere made no difference to Isherwood, since he was concerned with the normal operation of marine steam engines.

Isherwood could not ignore the fact that there was great power loss through absorption of heat from the steam by the cooler cylinder walls. Almost 20 years before, engineers in Cornwall, England, had insulated their engine and obtained 50 per cent greater efficiency than Isherwood did from the engine on the *Michigan*;[15] but in 1860, as in subsequent years, Isherwood insisted on taking practical factors into account. Throughout the Civil War he would have to build his engines to be simple and reliable, and if it meant a loss of economy through low pressures and no steam jacketing, this was the necessary price.

Many engineers protested that the Erie experiments were worthless because the steam pressure used in the engine was so low that it invalidated any test of expanded steam. A great expansion of steam would produce a corresponding drop in temperature, which would result in a sharp decline in steam pressure, so that in order to obtain a fairly high average pressure against the piston, the steam would have to enter the cylinder in an extremely compressed state. At Erie, Isherwood used steam at only a little over 20 pounds, which was not enough to make any engine operate effectively with steam expanded beyond a modest degree. However, in 1860, steam pressures were normally quite low, just as cylinders were usually unjacketed. Thirty pounds was about the maximum working pressure for boilers, although a federal law required a 50 per cent safety margin for boilers, meaning that, theoretically, they could be pushed to 45 pounds. Isherwood, however, decided that low boiler pressures were necessary for a test which would apply to ordinary marine engines, so that even 30 pounds of pressure was excessive for his experiments.

A third, though lesser reason for complaint was that Isherwood did not superheat his steam. This procedure was still quite new, and although it came into widespread use during the Civil War, it was not that common a naval engineering practice in 1860. By superheating, the steam from the boiler would be reheated, driving out moisture and increasing its temperature. In this state, when it expanded within the cylinder, steam would not condense so quickly and, therefore, would be better adapted for large measures of expansion. Of course, the greater efficiency of superheated steam was partially offset by the cost of the fuel used to provide this higher temperature.

The results of the Erie experiments on the *Michigan* showed Isherwood that his techniques, as well as his ideas, had great value for the engineering world. He saw his Erie work as "an additional admonition of the extreme unsafety of depending upon inference in the physical sciences." Only experimental results, obtained under carefully controlled conditions, could be relied upon to solve scientific problems. He had gathered all the data possible at Erie; and what was more, he published all that he observed, presenting his entire experimental procedure step by step, describing his method of calculation, and carefully stating the physical laws on which he based his results. The inductive process, he concluded, was the only way to derive scientific laws; and for his day, this was a practice rarely observed with true fidelity.[16]

His experiments drew flattering attention and critical acclaim abroad as well as at home. Professor Rankine, the world renowned expert on thermodynamics, read a paper, on February 5, 1862, to his professional associates at Glasgow, basing his remarks on Isherwood's experiments at Erie. Rankine treated Isherwood's work with respect, although questioning Isherwood's assertion that steam would condense through expansion *per se*.

To the scientific world, Isherwood appeared as a "bold writer" who had come forth to challenge the accepted theories of three generations of engineers. Everywhere engines had been built on the theoretical principles of steam expansion, and it was universally believed that such engines were the most economical. Suddenly this young man, the *Scientific American* noted in its issue of May 4, 1861, comes forth with claims which "have somewhat startled the engineering world," since they "strike at the very root of opinions long and generally entertained."[17] The steam trials at Erie were the first of their kind ever published, his critics admitted, and the credit for whatever new light they cast belonged to Isherwood.

In later years, the techniques Isherwood utilized at Erie excited more admiration than his results, valuable as the latter were. Robert Henry Thurston, the leading American writer and expert in steam engineering of the late nineteenth century, considered Isherwood's work on the *Michigan* to be the first systematically conducted investigation into this phase of steam-engine efficiency. Isherwood, he asserted, was the first to attempt to determine by a planned and systematic method the law governing cylinder condensation in relation to the degree of steam expansion. It was the first of a number of such investigations by Isherwood; and such was their thoroughness that, in 1889, they still could, as Thurston wrote, "constitute the principal part of our data in this direction."[18]

As Engineer in Chief and then as Chief of his own bureau, Isherwood had much opportunity for experimentation, and he took full advantage of it. In 1860 he had made his first tentative experiments with superheated steam, but it was a series of extensive trials made between 1862 and 1876 on many warships which gave Isherwood the data he required to form a conclusion on the value of superheating. He recognized the theoretical gain in engine efficiency by heating and re-evaporating the steam coming from the boilers, but, characteristically, Isherwood was skeptical. It took extra fuel to provide the extra heat; and, in addition, the steam, in its superheated state, posed problems in engine operation. The high temperatures and pressures of the vapor in its superheated state tended to wear down the moving parts of engines. Lubrication of steam valves and pistons became difficult, and there were continual injuries to the cylinders. Although superheating steam produced a gain of 18 per cent in operating economy, Isherwood cautiously accepted its use, stipulating that the amount of temperature increase should be limited to 100 degrees.

By 1863, Isherwood had managed to persuade the Navy Department to sponsor a full-scale series of experiments in steam expansion, based on his work on the *Michigan*. With an appropriation of $20,000 granted in March, 1863, by Congress, Isherwood and Horatio Allen, president of the Novelty Iron Works, in New York city, began a program of testing which soon grew into a formal investigation of all aspects of the expansion problem. In early 1864 the Navy Department set up the necessary apparatus; and, in June, a commission of nine experts, headed by Allen, began the supervision of experiments. Of this board of experts, three represented the Franklin Institute; three, the National Academy of Sciences; and three, including Isherwood, the Navy Department. John Ericsson was asked to be a member of the commission, but he declined the invitation, pleading overwork. Although the experiments, conducted at Allen's plant by naval engineers, were never completed, the very fact that they were begun and continued for years under such august auspices was quite a feather in the cap of the man whose work had forced the recognition of steam expansion as a controversial issue.

Meanwhile, Isherwood was busy with his own experiments. Taking advantage of the hundreds of ships now operating in the Navy, he began large-scale tests on marine boilers, coal, engines, valve gears, steam pressures, cutoffs, saturated and superheated steam, and many other phases of naval engineering. His basic aim, as he said of one experiment, was "to treat the subject in an exhaustive manner, resolving in a purely practical way all the questions connected" with steam engineering. He was particularly satisfied that he made his tests always under conditions of actual practice, so that the results would be applicable to the problems the Navy would face in the future.[19]

While modifying and refining naval machinery, Isherwood also turned his thorough and inquiring mind toward new devices. In 1864 he supervised a board of engineers which grappled with the possibilities of "petroleum" as a substitute fuel for coal in naval steamers. Their initial results were promising, Isherwood told Welles, in a report dated November 28, 1864; but they were stopped for the moment by a ticklish problem. When exposed to the air of a confined space at summer temperature, he reported, oil produced a gas which became explosive when mixed with fresh air. Until a remedy appeared, there was little hope of driving naval steamers with this fuel.[20]

Isherwood was not satisfied with encouraging and supervising the great mass of experiments made by the Bureau of Steam Engineering throughout the war. They would be of greater value, he felt, if he could make them readily available to civilian and naval engineers alike throughout the country, if not the world. On February 7, 1863 he wrote to Welles asking that his bureau be authorized to publish a yearly volume which would give specifications and performance data on all available naval steamers. In addition, his "special experiments," such as those on boilers and steam expansion, should be collated, analyzed, and printed with illustrations. Although this would be "a work of considerable time and labor," requiring expert theoretical and practical knowledge, Isherwood was convinced it would be worthwhile for the department to sponsor. He felt it was virtually impossible to exaggerate the value of the information, both to the government and to "all persons engaged in the manufacture or use of steam machinery." He estimated that the cost for such a project would be $4,000 for the 2,000 copies of the report to be printed.[21] Welles cautiously approved the idea in principle, and asked Isherwood's opinion of the cost of a more modest annual volume which would publish only experimental information. Isherwood quickly replied that this would run about $3,000, which could come out of the contingent expense fund of his bureau, so that no congressional appropriation would be necessary.

Despite Isherwood's persuasive arguments, Welles decided that there were no department funds available for publishing the experiments. The Engineer in Chief then submitted his material to several publishers, but without success, as they were reluctant to take on a project which would obviously be "too unremunerative." Fortunately, several "eminent engineers" who valued Isherwood's work and wished to have his data and conclusions in printed form stepped in and offered to meet the publishing costs. With this backing, Isherwood was able to have the Franklin Institute, for many years the organ for semiofficial reports from naval engineers, publish his book. In 1863 a large, heavy volume, crammed with statistics and spiced with a forthright statement of engineering philosophy, appeared before the engineering world. Though based on the data collected by many naval engineers, *Experimental Researches in Steam Engineering* bore the unmistakable stamp of its author, Benjamin Isherwood.

The purpose of this book, the engineer explained, was to determine the comparative economic efficiency of steam used with different measures of expansion. He took the recorded performances of a number of screw warships, including the *Merrimack* and *Roanoake,* and calculated the costs of power produced by their machinery. On each of these vessels he made an exhaustive examination of the effect of steam condensation and the improvements afforded by steam-jacketing and superheating. After providing a meticulous analysis of boiler design and operation, calorific values of various coals, and the performances of the screw frigates, he turned to the machinery designed by John Ericsson for his monitor ironclads.

Ericsson, Isherwood decided, designed quite inefficient engines. The cost of the power for the *Monitor* and the *Passaic* was at least one-eighth more than that for engines of the "usual," or Isherwood, type because of excessive steam condensation within the cylinders. Ericsson had placed the two cylinders of his engines end to end, standing athwartships; and by having a common wall, they suffered great temperature changes within the cylinders. Although refusing to state categorically that Ericsson's engines were unfit for naval service, Isherwood left no room for doubt about the inferiority of the monitor engines.

In discussing the results of his boiler experiments, the Engineer in Chief explained why he had employed a principle which contradicted accepted practice. Most engineers of the 1860's designed their engines with small boilers and large cylinders, assuming that increased power would depend on a large piston with its long stroke. Isherwood observed that their engines failed to develop sufficient power because the small boilers could not supply enough steam for the big cylinders, so that the engines could never work up to their full capacity. Consequently, the average steam pressure on the piston throughout the stroke would be low. Since friction and back pressure against the piston were constant, relatively little power went through the propeller shaft, and the economic results were poor.

Isherwood decided to design his engines by reversing the normal practice, using large boilers which would assure sufficient steam for the smaller engines. As a result, his machinery was among the first that had boilers capable of supplying all the steam the engines could work off. In this way, engines could be driven at maximum speed for long periods of time, and there would also be sufficient steam when using a long cutoff, as Isherwood so strongly recommended.

As there was very little such reliable and practical information recorded anywhere, Isherwood considered this book to be invaluable.[22] For him it was more than a beacon to light the way for his colleagues, groping their uncertain way through the darkness of ignorance and folly. It was, in addition, a true labor of love and a measure of his devotion to his profession. Far into the nights, "after the labors of a long day," the Engineer in Chief had remained at his desk to pore over the maze of statistics which he gradually assembled into an intelligible whole. "It is the product of a weary hand," he admitted, "and [of] an attention overtasked by a great number of pressing and important duties whose requirements had to be immediately satisfied, and in which responsibility could neither be avoided nor divided."[23]

Isherwood still did not abandon hope for government support of the publication of his experimental data and analyses. Since he had sufficient material for subsequent volumes, he reasoned that the Navy Department might assist with their publication, and if not that, might at least subscribe for a sufficient number to guarantee publication. In May, 1864 he asked Welles if the department could subscribe for 300 copies of the second volume of *Experimental Researches*. He already had 320 copies spoken for at $10.00 a copy, a very high price made necessary by the elaborate diagrams and many statistical tables which would fill the book. The department could get a reduced rate if they placed a large order, Isherwood suggested; and for 300 copies, the price would be only $8.00 for each book. This price would just cover the cost of publication, he assured the Secretary. Only 1,000 copies were to be published, and no individual would collect any profit from the book, "the sole purpose being merely to disseminate the knowledge."[24]

Welles, impressed by the Engineer in Chief's assertion that these experiments "furnish, in general, of almost every problem in steam engineering a practical and sound solution...," authorized the subscription.[25] With this backing, Isherwood once more obtained the support of the Franklin Institute, which published volume two of *Experimental Researches* in 1865. This oversized, six-hundred-page book, similar in content to volume one, featured a large number of experiments on boilers, of which one of the more intriguing was an elaborate test with tobacco juice to see if it could prevent boiler scale.

Isherwood, in his preface, paid graceful tribute to Welles and Fox who had "sought by every means within their control" to advance engineering knowledge and to encourage the Engineer in Chief in his labors. The experiments had been so extensive, and consequently so costly, that were it not for the liberality of the Navy Department, the work could never have been done. For the past two years, Isherwood noted, he had spent "every moment he could snatch from his official duties," in collecting the test data, and now he had enough material for yet another volume.[26] He was especially proud of the fact that these experiments had not been controlled, laboratory tests under ideal conditions. He had no choice in the machinery tested; none of it had been designed for experimental purposes, and he had used whatever equipment he could get his hands on. Therefore, he concluded, his analyses and conclusions were of the greatest utility because they applied to practical, everyday engineering situations and could not be attacked on the basis that he had rigged the testing procedures to obtain his results.

By this time, Isherwood and his theories were familiar to engineers throughout the world. Reaction to his *Experimental Researches*, therefore, was no longer neutral or casual. English engineering journals, in particular, had strongly and continually criticized his machinery designs and theories, so their reaction to this book perhaps best illustrated the respect with which Isherwood was held, even by his professional enemies. Editors of the London *Engineer*, taking a hasty glance, pronounced it to be ponderous, so much so that they could only announce the fact that they had seen the book and would try to review it at a later date. Nevertheless, they did not hesitate to discuss the capabilities of the author. They concluded, in the June 8, 1866 issue, that he was not only a careful experimenter, but probably "the most precise and elegant American scientific writer now alive." This being so, it was a pity that he lacked the power of making correct deductions from given facts. "Many of his experiments are conducted under conditions which are to the last degree unpractical and objectionable," *The Engineer* complained; and yet despite all his unpardonable theories and techniques, *"Experimental Researches* constitutes the most important work on the modern steam engine which has recently issued from the press," thereby deserving serious attention.[27]

Important as Isherwood's scientific experimentation was for the development of steam engineering, it was not the major contribution he made to his profession. In his years of experimenting, Isherwood had shaped a philosophy of engineering that was to enrich his books and set a striking example for his contemporaries. At a time when neither he nor his professional associates had received formal training in research methods, Isherwood stood forth as the man who initiated a new era in steam engineering. By applying scientific methods and scientific apparatus to prove the relation of wastes in the steam engine to the limitations of thermodynamic theory, he became the first American engineer to "fix a settled principle of promoting professional knowledge and the solution of practical problems in engineering," according to R. H. Thurston.[28]

Behind this work lay his firm conviction that his experiments should endure as an imperishable contribution to man's knowledge, and that any less exalted aim would make experimentation pointless. "Whatever your investigations," he counseled, "leave nothing omitted or imperfectly done. In establishing conclusions, make your experiments so exact that if under the same conditions they should be repeated by some other experimenter a hundred years hence the same results would be confirmed." It was this unequivocal stand which won him the admiration and obedience of his assistants, and inspired one to assert, in later years, that "in thoroughness, exactness and ingenuity as an experimenter no man has ever excelled Mr. Isherwood."[29]

So convinced was Isherwood in the efficacy of the scientific method and the virtues of scientific knowledge that he viewed man's existence largely in these terms. His profession became for him a mighty instrument of power, far exceeding in scope any utilitarian benefits it might offer society. "The engineering arts which enable men to conquer nature," Isherwood reasoned in an article published in *Cassier's Magazine,* "enable them to also conquer one another, the highest development prevailing."[30]

Logically, Isherwood interpreted the Civil War as an issue created by industrialism. The agrarian South, unable to coexist economically with its industrialized neighbor, had no chance against the mechanical and manufacturing North. The latter was destined to emerge victorious, he believed, "for the highest development of engineering art and science represents the highest development of physical power." Britain, a nation of mechanics as well as of shopkeepers, had already gained world ascendancy through her industrial might; and her recognition of this basic source of strength would preclude her intervention in favor of the South. As Britain was bound to realize the inevitability of victory for northern industrialism, Isherwood never took seriously the threat of intervention. So far as he was concerned, the belligerent elements in England were overruled by wiser heads who did not wish to lose northern trade, money, and manufactures.[31] What Isherwood apparently overlooked was the possibility that Britain might wish to curb the economic and industrial potential of the United States by effecting a division which would hamper the development of an industrial competitor.

The most detailed expression of Isherwood's engineering philosophy appeared in his two volumes of *Experimental Researches.* Although he described these books as simply "collections of original engineering statistics with the general laws deduced from them," this was not to minimize their value, since he believed "science is nothing but a similar collection of statistics." All real knowledge which man possessed came from observation and comparison, and man's only teacher was comparison based on accurate observations. Therefore, he considered precise experimental data as the *sine qua non* for man's development. Unfortunately, many men would not recognize this fundamental truth, so they faultily commenced their quest for knowledge with "imaginary assumptions that can only end in a vain parade of misdirected skill."[32]

For Isherwood the inductive method stood alone as the way to achieve scientific truth. The engineer had to start with the particular fact and work gradually and painstakingly toward the general proposition. Scientific genius was that capacity for perceiving general principles from the chaotic mass of facts surrounding them, but the facts had to be there originally.[33]

To be sure, the deductive method was certainly cheaper and easier. However, he had no use for the product of such faulty methodology, branding it as the "sophistical nonsense of amateurs and would-be engineers who deem *a priori* argument sounder . . .than laborious experiment." The prime example of this misguided faith in deductive reasoning was the exaltation of Mariotte's law, a principle so simple and plausible that those both building and using steam machinery "cling to it with a prejudice amounting to fanaticism." The proponents of this principle were pathetic figures to Isherwood. They steadfastly refused to admit the existence of any results which went contrary to their law, and made their own knowledge and extent of observation the standard and the test of truth. In doing so, Isherwood observed, they mistook their own horizon for the bounds of the universe.

Instead of false knowledge based on opinion, Isherwood demanded truth from experimentation. "All sound knowledge," he insisted, "must be observational—purely experimental. . . ." Honest, sagacious experiment, repeated over and over under all the various conditions of actual practice, alone could lead the engineer to true knowledge. The trouble was that even when engineers were willing to experiment, they did it in a halfhearted way. "If every engineer had made only one *complete* experiment and placed it on record in full detail," Isherwood believed, "engineering would be far in advance of its present position."[34]

Experimentation would not only build toward the perception of general principles, it would also give practical value to existing scientific laws. By modifying abstract theoretical generalities in light of their actual working results, experimentation checked the evils of overgeneralization. The function of the engineer, Isherwood insisted, must be a purely practical one. He should refuse to accept "abstract and subtle speculations" and concentrate only on determining concrete facts, just as a pilot should avoid contenting himself "with a theoretical deduction of the laws of the ebbing and flowing of the sea," and, instead, should acquire a knowledge of the direction and force of its tides.[35]

The danger with hypothesizing was that it encouraged oversimplification. There was an inevitable complexity to scientific truths, Isherwood felt. No effect could have a single cause, nor one cause a single effect. Experimentation had to be necessarily arduous and thorough because no single experiment could stand by itself. "The object of our search," he declared, was "mean numerical values," to stand as a basis for establishing physical laws. Innumerable tests must be made to eliminate the influence of a single erratic result. Only by this accumulation of data could there be any hope for a consistent and well arranged system of physical science, and for a general world view which Isherwood described as "rational empiricism . . . the observation of facts confirmed by the operation of the intellect."[36]

The lure of the quick, easy method was often too great. As Isherwood viewed the engineering profession of his day, he recognized that it was so often in error because engineers neglected to collect facts and trace proximate causes. Instead, they strained to achieve a comprehensive principle whatever the cost. "Misled by a false ambition to grasp everything and predict *a fortiori,* truths have been missed that might otherwise have been seized," he lamented. His contemporaries all too often ignored the physical conditions of a practical problem because it would involve complexity and lack of neatness. Instead, engineers cheerfully sacrificed practical considerations "to the neatness and perspecuity, and to the necessities and narrow scope of a mathematical theory."[37]

So often an infatuation with mathematics misled a promising engineer. Pure theorizing, Isherwood argued, was largely a "mischievous element" in engineering, since "the utility of mathematics in the arts is extremely limited, and the principal advantage arising from its application is its prestige." Nothing annoyed him more than a pseudoscientific paper crammed with abstruse calculations to gild the lily of pure speculation. Engineering was an art, Isherwood insisted, not a speculative science. Its problems had to be solved by practical application and experimentation, not by the massing of impressive algebraic formulae to buttress conclusions which were preordained by the initial assumptions of the speculative writer.[38]

To dispel the false validity and authoritative aura of scientific speculation was his mission, Isherwood concluded. His was to be the role of the iconoclast. Thus his experimental researches were doubly valuable; not only could he enlarge man's knowledge, but also expose his ignorance. Correcting an error was often more valuable than a positive increase in knowledge, he maintained, and he would combat a flood of analogies, inferences, definitions, and hypotheses to assure that science rested simply on direct fact. He would rescue engineering "from the vagaries of empirics," and defend it against "a darkening of knowledge by words without understanding."[39]

He therefore made certain that his own work would not lack observed facts. Repetition and minutiae filled his books, but he refused to apologize for such meticulous detail, even if it resulted in prolixity. Conciseness and condensation were unthinkable if they resulted in the "omission of even the most trivial fact." Anyway, he argued, his material would be easier to comprehend if his presentation was occasionally repetitious and included frequent summaries.[40]

Most writers, he observed, went the other way and presented the reader with a *"mirabile dictu"* type of experiment where the results alone would appear. Even when they condescended to include the details of their tests and the processes employed, these writers would give such meager and contradictory information that it was worthless. As far as Isherwood was concerned, such engineering reports were mere advertisements, published only with the hope of promoting new inventions. The motivation behind these reports was private gain; there was little, if any, indication of disinterested scientific aims.

If "men cannot make a world from their own conceptions," as Isherwood maintained, they were certainly willing to try. The pervasiveness of Mariotte's law was sufficient evidence of that. So long had it been "an undisputed article in the creed of engineering," that Isherwood recognized the dangers he faced in daring to overthrow this idol. Yet Mariotte's law was not a physical law. If it were, it would be an expression for the application to steam of generalized physical facts, which it obviously was not. In steam engineering it was just a mere hypothesis, a conjecture whose only real value was in giving direction to experiments. However, it had become ingrained as engineering dogma, and its acceptance as physical truth was "worse than nonsense," according to Isherwood, since it led engineers astray. Drawing on the traditions of scientific progress, Isherwood saw the necessity to disprove this pernicious theory, since "it is often as important to overthrow as to establish, and the reformers of philosophy must be stern iconoclasts." His was the task of combatting both those deluded engineers who "cherish a fallacy from habit," and the general *"vis inertia* of man."[41]

If Isherwood chose to destroy cherished beliefs, no matter how fallacious, it was reasonable to expect resistance to his efforts. He was especially vulnerable, as he not only tried to tear down the favorite idols, but also set up his own alternative theories in their place. As Chief of the Bureau of Steam Engineering, he could both design and have built a large part of American naval machinery on his principles, so the performance of these vessels would presumably prove the validity of his theories.

That his naval machinery had to be built for specific purposes and to withstand unusual abuse was generally ignored. The only thing that mattered to his critics was the performance data of his machinery compared to any other. At least they often challenged him on his own ground, since he had always argued for working conditions and practical operation as the true tests of any engineering theory. The difficulty, as he had already observed, lay in his critics' refusal to consider dispassionately either his published experimental data or the performance of machinery built on his theories. It was sufficient that he had dared to challenge a sacred dogma; for this he had to be wrong, and had to suffer the consequences of his error.

Fortunately, Isherwood had never been a man to curry favor or require praise. His indifference to popular opinion now stood him in good stead. In an era noted for the flourishing art of personal abuse, Isherwood drew his full portion. In the daily press, in professional journals both at home and abroad, in frequent pamphlets, in Congress, and in the Navy Department itself, a continuing flood of criticism and vilification swept around the Engineer in Chief as he worked unceasingly to build the steam Navy and encourage the development of engineering.

So convinced was the Engineer in Chief in his cause that he could usually shrug off his critics. They had "violently impugned" his facts and his deductions, but "apparently with more of wounded vanity and alarmed interest than of philosophic spirit," he noted. Their personal censures of him, in which they indulged "with but little delicacy and no propriety," he dismissed as mere "petty annoyances of life which the experienced expect and the wise disregard."

None of his critics had shown his reasoning to be faulty, so he refused to reply to their assaults. Their I-do-not-believe-it attitude, he felt, was not enough to warrant a defense on his part, since the only way in which they could really challenge him would be to produce different results from their own experiments. This, he reasoned, was an impossibility, as his "purely practical plan of investigation" had eliminated the chance for error and had rendered his position "absolutely impregnable." Thus his principles might be challenged, but never "the truth of the grand results" of his experiments, which were so important that it was only a matter of time before they would be universally accepted. "Hence my indifference to the attacks of criticism," he comfortably noted, "arose from a deep-rooted conviction of their utter inability to make an impression on the massive stronghold of facts I had erected."[42]

He felt magnanimous enough to take pity on his enemies. Unwilling to humiliate them needlessly, he felt that "the weakness of human nature ... rendered charity a justice." If he destroyed their cherished theories in too precipitate a manner, he would really be destroying his critics, because it would "be taking away their foundations of knowledge," leaving them pathetically naked and shamed in the eyes of science. After all, Isherwood said tolerantly, "Few have the genius to discover truth, and not many have the good sense to distinguish it when discovered."[43]

The proper course, then, was not to destroy his opponents, but to persuade them of the truth. Of course, toleration of new ideas is never automatic or easy. It is a normal if regrettable tendency, he admitted, for men to enshrine theories which have merely existed for a time without being refuted or challenged, and then reject all facts inconsistent with these theories. Galileo was not believed, despite the evidence he presented, merely because his data was inconsistent with existing astronomical hypotheses. Isherwood's detractors should thus cast off their narrow-mindedness and admit the truth as revealed by this Galileo of nineteenth-century steam engineering. "We must possess ourselves of that intellectual liberality which accepts without reluctance whatever is demonstrated to be true, however contrary to our previous belief," Isherwood admonished them, "and dismiss without regret whatever is shown to be false, however cherished the fallacy may have been."[44]

The Engineer in Chief was too optimistic. His unshakable conviction in his own beliefs made him appear insufferably patronizing rather than persuasive. Men were not so flexible as Isherwood had wished, and too often they greeted his philosophic pronouncements with derision. Too many personal and financial factors were at stake for them to cast aside their beliefs and accept the word of Isherwood. As many engineers resentfully felt, he was simply an obstructionist who had selfishly seized on his key position in the Navy Department to force the acceptance of his own theories and designs while viciously resisting the use of any others. Self-interest, as Isherwood would realize, was far too durable to melt away in the face of the truth.

# V. The Lawyer and the Engineer

Of all the critics who badgered Isherwood there was none more persevering, virulent, infuriating, and effective than Edward Nicoll Dickerson. His attacks began before 1860 and persisted almost until the end of Isherwood's tenure as engineer in chief. His motivation may have been purely professional, but such were Dickerson's energy and ingenuity that he turned his clash with Isherwood into a *cause célèbre* in the public press and in Congress, as well as among the engineering profession. Although Isherwood withstood the repeated assaults of Dickerson, he did not escape unscathed. His professional and personal reputation was continually in question throughout his years as bureau Chief, largely because of Dickerson's efforts.

Edward Dickerson, in retrospect, appears ludicrous, worthy of notice only by his providing color and spice to the pages of history. In his day, however, he presented a far different picture. Six feet, three inches tall, he towered over his contemporaries, no doubt intimidating many by his imposing figure alone. In keeping with his impressive size was his dignified manner and determined stride, which, for many observers, marked him as a man of decision, purpose, and massive self-confidence.

Dickerson received a more impressive education than Isherwood, his interest in science having been stimulated by Joseph Henry, while at The College of New Jersey, in Princeton. It was Henry who turned Dickerson to a lifetime study of mechanics which would lead him finally into the field of steam engineering and to his bitter struggle with Isherwood. Only two years younger than the Engineer in Chief, Dickerson also had to learn his engineering by working directly with engines; and like Isherwood, he began his practical education on the railroads, becoming the first man to run a locomotive between Paterson and Jersey City, New Jersey. Instead of becoming a practicing engineer, however, Dickerson turned to the study of law with such success that by the 1850's he was reputed to be the outstanding authority on patent law in the United States.

Despite this choice of profession his fascination with engines never left him, for he retained engineering as an avocation to which he could apply his fertile and clever mind.

In the late 1850's he collaborated with the inventor Frederick E. Sickels to display his engineering qualifications to the general public, by winning a contract to build a new pump engine for the city waterworks in Detroit. After four years of innumerable delays in construction and continual modification in the engine design, the engine was completed, but it ran so badly it could never be brought to a full trial run.

Undaunted, Dickerson and Sickels had already turned to designing machinery for steamers. On April 3, 1858 they obtained a contract from Secretary of the Navy Isaac Toucey to build the machinery for the sloop of war *Pensacola*. Toucey had been quite reluctant to grant them this favor, since neither man had ever designed a working marine engine. Moreover, Dickerson had become intrigued with the possibilities of steam expansion and was convinced, in accordance with Mariotte's law, that he could design steam engines using high measures of expansion to produce great economies in operation. By carrying this generally accepted theory to an extreme, Dickerson had alienated the engineering profession, which, in any case, resented the appearance of a lawyer in their domain. Engineer in Chief Daniel Martin led the opposition of naval engineers to Dickerson's plan for the *Pensacola,* and received abundant support from civilian engineers who made Dickerson and Sickels "the common topic of ridicule in engineering circles."

However, Dickerson was as resourceful as he was visionary. He carefully cultivated his friendship with Senator Stephen Mallory, chairman of the Senate Naval Affairs Committee, and with D. L. Yulee, also a Florida senator and an influential member of the same committee. With both men Dickerson became "intimately connected, socially and politically," and for this he drew ample reward. Despite the opposition of John Lenthall and the engineers, Secretary Toucey finally capitulated to the political influence of the senators.

Once awarded the contract, Dickerson and Sickels then attempted to build their engine. During construction they made so many and such radical changes to their machinery that the *Pensacola*'s hull had to be redesigned to accommodate their structure. It took two years and nine months to complete the machinery—about three times longer than normal. The total cost was $328,460, and even then the machinery proved too heavy and bulky for the vessel, forcing such a reduction in the amount of coal the steamer could carry that her steaming endurance was cut almost in half.[1]

When the *Pensacola* finally had her trial in December, 1861, she developed 8.8 knots at the start; but during the trip from the Washington Navy Yard to Key West, Florida, she began to behave erratically and could average little better than 5 knots. After a friend of Dickerson's made repairs on the vessel, the steamer made 7 knots, which was still far below the requirements of the Navy Department. Despite these inadequacies, the Navy Department needed the *Pensacola* sufficiently to order the steamer to join Admiral Farragut's fleet on the lower Mississippi River. Here the machinery gave out entirely. Farragut, in disgust, first moved her by sail power and finally had her towed up and down the river as a floating battery.

In early 1863 four Navy Department engineers surveyed the *Pensacola*'s machinery and unanimously pronounced it an unmitigated failure. Dickerson then pressured Congress for an investigation by civilian engineers who presumably would not be prejudiced against him. Even here he had little satisfaction, for of the eight civilian experts who examined the *Pensacola,* only two failed to condemn the machinery. Isherwood, who finally removed Dickerson's machinery and replaced it with his own engines, was understandably annoyed with the waste of time, money, and effort resulting from the department's humoring of Dickerson's eccentricities. Writing to Welles on February 17, 1863, he branded the *Pensacola*'s original engines as unreliable, wretched, entirely useless, and absurd creations of ignorance. Dickerson's engineering theories, he asserted, were nothing more than the "conclusions of charlatanism" and should never be dignified by a comparison with the "matured productions of engineering skill" emanating from the steam bureau.[2]

Dickerson's and Sickels' unique designs had caused him far too much trouble already, Isherwood decided. Not only had the *Pensacola* become a victim of their visionary schemes, but other vessels had suffered as well. The steamer *Richmond* had been fitted with the same Sickels patent cutoff-valve gear used on the *Pensacola* and, consequently, had become the slowest steamer in the Navy. Working in her engine room was a nightmare for her Chief Engineer, J. H. Warner, who wrote to his Chief:

I sincerely hope it will never be my fortune to have charge of such an unnecessary complication of machinery, in the shape of a marine engine. It is excessively noisy, and can be compared only to a cotton-mill with power looms.[3]

Dickerson understandably took a different view of the tribulations of his machinery. Like Alban Stimers, he simply could not envision the possibility of his own error or inability. The logical explanation, therefore, was that the Navy Department, mainly through the machinations of its Engineer in Chief, was persecuting him. Ever since Dickerson had begun to design machinery for the Navy, Isherwood had opposed him, undoubtedly, he believed, through professional jealousy, if not through personal malice; and now that Isherwood had risen to command the steam bureau, he had emerged as the patent lawyer's *bête noire*. His own duty, Dickerson decided, was quite clear. He had to destroy the pernicious influence of Isherwood for the good of the service and of the nation, let alone his own welfare. To this end, Edward Dickerson applied all the skill, imagination, and energy at his command.

Dickerson began his campaign against Isherwood in early 1861, when he learned of Welles's choice for engineer in chief. Dickerson later asserted that he had been primarily responsible for the congressional opposition to Isherwood's nomination. Angered by the restriction of the position to naval engineers—a tactic which he attributed to Isherwood—Dickerson had applied his persuasive talents to profitable effect. That Isherwood finally received confirmation the lawyer regarded as a regrettable error, mainly because of the influence of Welles, for whom Dickerson exhibited no great respect thereafter.

Nevertheless, Dickerson did not totally discount the use of the Secretary of the Navy as a means to thwart Isherwood. On January 6, 1863 he addressed a public letter to Welles, reiterating arguments which he had published in *The New York Times* several days before. Printed as a pamphlet and entitled *The Steam Navy of the United States: Its Past, Present, and Future,* this letter set forth a general attack on Isherwood's theories, as expressed in *Engineering Precedents;* on the poor performance of steamers built with Isherwood engines; and on the questionable relations between Isherwood and the naval engineers Martin and Sewell, whose boiler and condenser Isherwood insisted on using.

Dickerson, renowned for his courtroom histrionics, displayed his flamboyant and provocative style in his letter to the Secretary. To a layman such as Welles, the lawyer's engineering arguments were difficult to challenge, but Dickerson took no chances. In a covering letter to his pamphlet, he maintained that "Isherwood's absurdities" were so obvious that no "special science" was needed to perceive them. If Welles, for some incomprehensible reason, still insisted on treating them as scientific deductions rather than as errors, then he should consult some eminent scientist in order to correct his misapprehension. Professor Joseph Henry, Dickerson thoughtfully suggested, would be a fine authority.[4]

Dickerson began his pamphlet by asserting that ignorance had ruled the engineering department of the Navy since the inception of the new administration, for Isherwood was simply an "engine-driver," whose menial occupation never should be confused with that of an engineer. In daring to question Mariotte's law, Isherwood had shown himself to be an upstart who slandered the work of the great steam engineers and scientists, including Watt. As a result, the United States Navy now suffered from the depredations of Confederate raiders because the Union steamers were too slow, for which the blame had to fall on the "stupidity and ignorance" within the bureaus of Lenthall and Isherwood.[5]

By "active political service," the lawyer continued, Isherwood had insinuated himself into his present position. He had then proved his lack of engineering ability by producing *Engineering Precedents,* in which "every material proposition, *without* exception, is demonstrably false."[6] The fact that it was sumptuously bound in blue cloth and lettered in gilt could not hide the fact that its contents were worthless. The Engineer in Chief, whose proclivity for epaulettes and gold lace seemed incongruous for an "engine-driver," was as hollow and superficial as his book. Isherwood had risen by devious methods from his "respectable but humble position of an engine-driver," and now had put on airs and flaunted a "presumptuous dogma of his own, unsupported by any man of recognized knowledge in the world." He had shut the door against all outside influence and ability and had grasped the responsibility for the entire steam Navy.[7]

Someone, Dickerson pleaded, must save the country from this man who was wasting hundreds of millions of dollars. Out of pure, disinterested patriotism, and despite the knowledge that he was incurring the enmity of the naval engineers and all those preoccupied in protecting their vested interests, Dickerson offered himself as the champion of truth and justice.

The Navy Department refused to heed Dickerson's warning. Nevertheless, his call to arms won him a sympathetic audience among engineers and with the many groups who opposed the policies if not the accomplishments of the Engineer in Chief. Places hostile to the administration, such as New York city, were particularly receptive to his criticisms of the Navy's conduct of the war and provided Dickerson with a forum for his ideas. His letters to *The New York Times,* signed "Vindex," in keeping with the general use of pseudonyms, sparked a lively debate over the merits of Isherwood's engineering. Although the Engineer in Chief found occasional support from such friends as Henry Waterman ("Crucis") and Robert H. Thurston ("Thor"), the volume, if not the weight of argument was against him. John Ericsson and his assistant, Isaac Newton, willingly abetted Dickerson in disparaging Isherwood, and with virtually every newspaper in New York city hostile to the Navy Department, Dickerson enjoyed a responsive and wide-spread audience.

While in the midst of his busy attacks on the steam bureau, Dickerson found an ideal opportunity to confront Isherwood directly and thereby ridicule and destroy the professional reputation of the Engineer in Chief. In a trial held before the Supreme Court of the District of Columbia, the question of the efficiency of the Sickels cutoff had risen between the contending parties. Called as an expert witness, Isherwood gave testimony which was largely based on his experiments and his conclusions about the use of cutoffs with steam expansion.

Dickerson, asked by the plaintiff to rush to Washington and refute Isherwood's testimony, seized this chance to exercise his considerable courtroom talents against his enemy. After a display of sarcasm and withering logic which, he declared, left Isherwood "writhing under the cross-examination which was unmasking his villainies,"[8] Dickerson's impassioned summation before the jury won the case for the plaintiff. So eloquent was the lawyer in his closing address to the jury that he decided his remarks should be preserved and publicized. On January 1, 1864 he therefore issued a lengthy pamphlet entitled *The Navy of the United States: An Exposure of Its Conditions, and the Causes of Its Failure.* In its eighty pages the vilification of Isherwood undoubtedly reached its zenith. The Navy Department could no longer afford to ignore Edward Dickerson.

Dickerson's second pamphlet was essentially an expansion of his first, but also it included lengthy excerpts from the cross-examination of Isherwood. Dickerson modestly minimized his skill in persuading the jury that Isherwood's testimony was worthless, asserting that "it would be mere child's work to expose his ignorance and his corruption; for both are of so conspicuous a kind as to be obvious to the dullest vision."[9] It was a tragedy, he lamented, that such a man should be allowed to attain such a responsible position, let alone receive the continued support of the Navy Department. Because of Welles, Dickerson's efforts in the newspapers and before Congress to prevent the confirmation of Isherwood's nomination had been thwarted, which, he declared, had allowed that irresponsible engineer to "enter upon his career of destruction and fraud with the entire approval of his superiors."[10]

Once Isherwood had begun to produce his useless engines, Dickerson continued, the department had compounded its guilt by appointing a "white-washing committee" to inspect Isherwood's machinery. Since this committee was made up of contractors building Isherwood's machinery, it would not have been too surprising, Dickerson felt, for them to approve the Engineer in Chief's work. Nevertheless, by some quirk of human nature, they had been honest enough to censure Isherwood's engines, at which point their report was "accidentally mislaid," and Welles could only remember that it had been quite favorable to Isherwood.

As Chief of the steam bureau, the lawyer insisted, Isherwood had utterly demoralized the engineering corps. No one dared to oppose him, yet his theories and techniques were the despair of any intelligent engineer. The nadir of Isherwood's career arrived when he had the temerity to publish *Experimental Researches,* a piece of "barefaced fraud, which has no precedent in the history of the world in this department of swindling." Isherwood knowingly made false entries in his material in order to "force the balances" and prove his specious theories.[11] "Teeming with fraud," *Experimental Researches* was of "no possible interest to any man of science on earth, except the metaphysician and moral philosopher engaged in defining the shadowy boundary which separates the dark regions of moral depravity from the scarcely less obscure limits of intellectual insanity, to whom it would afford abundant material for researches in either direction."[12]

Isherwood's only excuse for publishing this abomination, Dickerson decided, was that his "moral obliquity" was so intense that he failed to recognize a fraud of his own making. In any case, the Engineer in Chief had coerced contractors into providing financial support for publication, and then had forced every naval "engine-driver" to buy it at ten dollars a copy. Without this engineering gospel according to Isherwood, the lawyer insisted, they could never hope to advance in their profession.

In the more mundane areas of corruption, Isherwood was equally at home, Dickerson observed. He had forced the Navy to use the exposed, expensive vertical water-tube boiler for which Daniel Martin fraudulently claimed patent rights actually belonging to a man named Montgomery. Disregarding legal decisions about the patent ownership, Isherwood made contractors pay royalties to Martin, who presumably split the proceeds with the Engineer in Chief. Much the same thing was done with the Sewell condenser.

A more direct source of funds for Isherwood was selling copies of engine drawings to contractors, who, anxious to avoid the delays which would otherwise result, willingly paid out $1,000 per copy. This practice was good for at least $20,000 to $30,000, Dickerson estimated, on the basis of the large number of engines under construction. The advantage of this particular scheme was that it was not legally fraudulent, since the contractors, rather than the government, would be paying out the money.[13]

These financial arrangements proved that Isherwood was "the right man in the wrong place." With the ability to make so much money, he should be secretary of the treasury, the patent lawyer wryly noted.[14] Probing into the source of Isherwood's wealth, Dickerson discovered that only a few years before the engineer had been poor. His widowed mother had been destitute and dependent on her son for support. On the small salary paid to a naval engineer, how had Isherwood suddenly accumulated the fortune about which he now openly boasted? For Dickerson there could be only one answer.

It was quite true that Isherwood had suddenly acquired immense wealth. Dickerson asserted that the engineer's mother had recently purchased $8,000 worth of New York city stocks at a 6 per cent premium, a practice in which she had often indulged during the last 2 years. This also was undoubtedly true, but his accusation that corrupt transactions provided the flow of money was another matter.

By his marriage to a young widow whose parents and previous husband both had apparently been well-to-do, Isherwood may well have obtained at least a small initial capital. According to his granddaughter, Mrs. Madeleine Kerwin, he rapidly augmented his funds by speculating in gold during the Civil War. He apparently would receive advance news of Union military failures or successes and then telegraph his mother who would buy or sell on the New York money market accordingly. Taking advantage of the rapid fluctuations in gold and utilizing "inside information," Isherwood had the opportunity to make huge profits by quick trading. His mother, a shrewd, strong-willed woman, played an active role in this operation. Her son apparently supplied the necessary tips, and she did the rest. Little wonder that Isherwood, in later years, could honestly say he owed all he had to his mother.

The weakness of Dickerson's attack ironically lay in his sources of strength. He was an eloquent, passionate, and highly educated man whose persuasive abilities had become sharpened by his years of practice as a successful attorney. Such was the power of his presentation, however, that he allowed himself to run to excess, swept up by the intoxication of his words and blind to the extravagance of his claims. He saw a world painted in colors too vivid for others to comprehend, and he dismissed moderation as lack of conviction.

As his grand opponent, Isherwood had to appear as the embodiment of evil, but in so attacking the Engineer in Chief, Dickerson lost the respect and credulity of much of his public. Isherwood, he claimed, "never has laid hands on a piece of machinery in his life." The Engineer in Chief, before "gaining his position by intrigue and brazen effrontery," had been a mere "penny-a-liner by trade, making a scanty living by picking up news in Washington for the newspapers, and writing letters on both sides of the questions of the day." He was a clown, a base fool, "an ignorant charlatan, having climbed into place and station by the practice of his base devices."[15]

Entertaining as they were, these descriptions were obviously not accurate, and they served to cast doubt on the lawyer's veracity. The 1860's were years of flamboyance, when the press was riddled with exaggerated and, at times, nonsensical claims; but even for those days, Dickerson was intemperate, often drowning his effectiveness in a flood of abuse.

Gideon Welles, for one, was disgusted. Dickerson's statement before the court in Washington was "a tissue of the vilest misrepresentations and fabrications that could well be gathered together," and merely served to provide fuel to the fires of administration critics. This "morbid love of slander and defamation," Welles observed, was "pitiable to witness," since the partisan press in New York would seize upon the most vicious assertion and publish it as the truth. Isherwood, the object of Dickerson's "reckless assault," may have made errors, Welles admitted, but the abuse he had received could not be justified or excused. Unprincipled congressional opponents of the Navy Department, such as John P. Hale and Henry Winter Davis, would use Isherwood as an excuse to condemn the entire department, while venting "their spite and malignity" on the Engineer in Chief. "There is an evident wish," Welles noted in his diary, "that Isherwood should be considered and treated as a rogue and a dishonest man, unless he can prove himself otherwise. Truth is not wanted, unless it is against him and the Department."[16]

Though no "partisan in engineer's difficulties," and despite his recognizing that the current variety of theories indicated that no one man had a monopoly on the truth, Welles still could not treat Dickerson impartially. There may have been some validity to Dickerson's theories on steam expansion, but his "misrepresentations and gross abuse of Isherwood" forced Welles to dismiss the patent lawyer's assertions as the rantings of a fanatic.[17]

Even Isherwood's critics admitted that Dickerson often harmed their cause. The *Army and Navy Journal,* bitterly hostile towards the Engineer in Chief, said, on December 26, 1863, that Dickerson's charges against Isherwood were sufficiently serious that they should be investigated, but that the lawyer's "fierce denunciation and hair-splitting logic, so common to lawyers, should be accepted with caution."[18] The Engineer in Chief might be guilty of professional errors, but slandering his personal character was not the proper way to judge him.

Although some observers considered Dickerson's courtroom summary "as an example of eloquent invective ... worthy of classification with the famous oration of Cataline," Isherwood was relatively unmoved. Preoccupied with his official duties, he found little time to bother over Dickerson's attack. Perhaps, as his friend Frank Bennett later recalled, Isherwood was little disturbed because he had been "too long and too prominent in public life to be supersensitive to criticism."[19]

Dickerson, however, did not share Isherwood's usual stoicism. So adept at dealing out abuse, the lawyer was not so able to receive it. Writing to *The New York Times* shortly after the appearance of his *Navy of the United States,* he complained of being exposed to "the most merciless assaults by those whose villainies or incapacity I am exposing." In return for his unselfish service in making a "thankless and laborious attempt to reform those monstrous abuses in the navy," he had received "no aid from any quarter," as the possibility of his failure discouraged any would-be supporters from braving the ultimate vengeance of the Navy Department.[20]

The attitude of Gustavus Fox particularly annoyed Dickerson. The Assistant Secretary of the Navy, deciding that the patent lawyer was more of a fool than a rogue, amused himself by distributing a pamphlet entitled *Uncle Samuel's Whistle and What it Costs,* a satirical discussion of Dickerson's questionable reputation as a steam engineer. Studded with numerous cartoons vividly demonstrating the arduous and dangerous duties of any engineer who ventured into the engine rooms of Dickerson's vessels, the pamphlet cleverly lampooned the engineering concepts of the patent lawyer. Distributed to influential congressmen, it went far towards making Dickerson appear a ridiculous and utterly impractical visionary.

Furious that his opponents would employ his own techniques, Dickerson published an indignant open letter to Fox, accusing him of responsibility for this "scurrilous" pamphlet. Three years earlier, when he had begun "to prevent the ruin of an entire navy," Dickerson had forseen that "the reckless and desperate man, whose schemes of profit I was endeavoring to thwart, would resort to the most desperate means for averting the consequences of the exposure." He had prepared, Dickerson asserted, "for every attack, even on my life," but he had not imagined that Isherwood could so entangle Fox in his intrigues as to use him as an instrument of revenge. The Navy Department, he declared, had employed an army of "detectives and pimps" to dog his footsteps and search for any vulnerable chink in his pristine moral and intellectual armor. Moreover, Fox had sadly cheapened himself by participating in character assassination.[21]

The Navy Department, Dickerson decided, had chosen to discredit him personally rather than disprove the facts he presented. Fox had "disgraced the office he holds" by trying to get New York papers to publish a "gross, false, personal libel," while Senator Grimes, of the Naval Affairs Committee, had been misled into offering another "false and libelous attack" to a New York paper. When the paper quite properly refused to print it, Fox tried to put on pressure to have it accepted, Dickerson maintained. The ringleaders of the conspiracy against him were Isherwood and Fox who, he asserted, were assuring both the President and the Congress that Dickerson's attacks on the department were for the selfish and ignoble end of getting Sickels' cutoff used on naval steamers. Writing an open letter to the Naval Affairs Committee of the House of Representatives, Dickerson protested against these tactics. While he was spending "so much thought and labor . . . in the hope of doing some good to our country," all his efforts were being ruthlessly sacrificed "in order to gratify the vengeance of one man" and to prevent the disclosure of Isherwood's ignorance and villainy.[22]

Dickerson was not so friendless as he often maintained. His enthusiasm and eloquence were often persuasive and, in one instance, profitable. Despite his failures with the *Pensacola*, Dickerson still determined to build warships which would reveal the inadequacies of Isherwood's engineering theories. By 1863, he had enlisted the aid of a wealthy and reputable New York merchant and shipowner, Paul S. Forbes, and it was Forbes's financial assistance and reputation which permitted Dickerson to attain his ends. Without the aid of Forbes, Dickerson never would have been taken seriously in his later offers to produce warships. With Forbes's backing, however, the patent lawyer had powerful influence to wield against both the Congress and the administration.

Forbes fully accepted Dickerson's engineering theories to the extent that at one point he reported enthusiastically to Fox that Dickerson had devised the last word in naval rams—a movable one which would be suspended from the bow of a warship and, by being connected to the engines, would sink an enemy vessel by trip-hammer action.

Although Forbes disapproved of the lawyer's tactics, he agreed that the Engineer in Chief should be removed for the good of the Navy. Writing to Fox in June, 1863, he warned the Assistant Secretary against both the theories and the practices of the Engineer in Chief. Isherwood's "invidious influence," he believed, seriously endangered the best interests of the department. Although Dickerson could protect himself against any personal attack made by the bureau Chief, there was little hope that vessels built on the lawyer's principles would ever obtain a fair consideration from the Navy Department so long as Isherwood remained.[23]

In February, 1863, Forbes offered to build for the Navy Department a sixteen-knot steam frigate in answer to the Secretary of the Navy's call for fast ocean-going cruisers. As this vessel would, with others, come into competition with similar ships powered by Isherwood's machinery, the approval by the department of Forbes's offer heralded a bitter battle.

Forbes proposed to build this cruiser, named the *Idaho,* in four to five months and at a price of only $600,000. The hull would be constructed by the famous designer Henry Steers, and the machinery, of Dickerson's design, would be erected at the Morgan Iron Works, in New York city. Since he did not intend to build this ship merely to make a large profit, Forbes explained, he was willing to sell her to the government at cost. However, knowing that the department did not conduct its business in such a manner, he had to set an initial price on which the government could make progress payments. This amount, he assured Fox, was not at all exorbitant, and he would rather lose a considerable sum of money than fail to deliver the ship, such was his desire to aid the nation.

Despite the fact that Dickerson's questionable abilities as an engineer were involved, Forbes had little difficulty in obtaining the contract for the *Idaho* in May, 1863. Isherwood was sufficiently unpopular in Congress and in the Navy Department to elicit a certain amount of sympathy for Dickerson. Moreover, Isherwood had cruiser machinery designs of his own which were so startling that the department was quite willing to consider the plans of any outsider, especially when backed by a man of Forbes's reputation.[24]

Once the vessel was begun, Forbes encountered the same problems as did other constructors during the Civil War. His confident promise to finish the *Idaho* within five months soon appeared pathetically optimistic, as time went by and construction of the ship progressed slowly, beset by problems in obtaining materials and by high wages and scarcity of labor. Not the least of Forbes's worries were the progress payments he had confidently expected from the Navy Department. With Isherwood and Lenthall in a position to supervise the construction and withhold payments if their specifications were not met, Forbes, by August, 1864, was asking for Fox's aid.

To complicate the problem, Dickerson's bitter assaults on the Navy Department, especially on Fox, had seriously jeopardized any good will the department still had for Forbes. On August 1, Forbes complained to the shipbuilder George Quintard, "I am afraid Dickerson's letter has thrown all of our fat in the fire." Unable to stop the lawyer's incendiary attacks from being published in *The Times*, Forbes asked Quintard to help him placate Isherwood and Lenthall, as well as Fox.[25]

By December, Forbes's situation had not improved, and he was in embarrassing financial straits. "Forbes is here in great trouble," wrote Congressman John Griswold from Washington to his friend John Ericsson. "He proposes to petition Congress for relief from his contract and stands just as much chance of success as of getting to heaven in a balloon." Griswold related that Alexander Rice, chairman of the House Naval Affairs Committee, believed the only way Forbes would get relief would be to get elected to Congress and find a place on the Naval Committee where he could protect his interests. "The more I think of our escape," Griswold concluded, "the more I am astounded."[26]

# VI. Trials and Tribulations

While the *Idaho* was under construction and Forbes was struggling to obtain the payments, Dickerson had embarked on yet another project which proved to be of much greater interest, and which largely settled the issue between the lawyer and the Engineer in Chief. When the *Sassacus* class of wooden-hulled, double-ender gunboats were contracted for in the autumn of 1862, Dickerson had proposed to the Navy Department that he be allowed to put machinery of his design in one of the vessels in order to provide a real comparison with Isherwood's engines. After a period of heated discussion, the department finally agreed to this proposal, and the *Algonquin* was placed under construction in March, 1863, at the Brooklyn Navy Yard, with the agreement that Dickerson's machinery would be placed in the vessel. Paul Forbes, not yet disillusioned by his experiences with the *Idaho*, made the machinery contract with the government on March 20 and promised delivery on September 20. He then subcontracted the construction of the Dickerson engines to the Providence Steam Engine Company in Rhode Island.

The machinery was finally completed and installed in June, 1865, and featured a boiler designed by Dickerson, a surface condenser of his design, Sickels' cutoff, and Dickerson's single inclined engine with a cylinder 48 inches in diameter and a 10-foot piston stroke.

The vessel chosen to compete with the *Algonquin* was the *Winooski*, launched in July, 1863, and similar in all respects except that her boilers were of the Martin type, and the engines were Isherwood's, having a cylinder 58 inches in diameter with a 105-inch stroke. The main difference was that Isherwood's boilers furnished a great deal more steam, in accordance with his principles, and Dickerson's engines were to be operated on the basis of extreme steam expansion using a very short cutoff.

By the time the two vessels were ready for their competitive trials, an unusual amount of public interest had settled on the Isherwood-Dickerson controversy. Isherwood's refusal to answer the lawyer's public attacks, and the persuasive eloquence of Dickerson resulted in a general support of Dickerson by the press and the public. In addition, the end of the Civil War, as Frank Bennett has observed, permitted more time for personal jealousy and dislike to assert themselves:

> All the disappointed contractors, all the naval officers whose official toes had been trodden upon, and all the politicians whose favorites had failed to retain the positions for which they were unfit, banded together and for two or three years filled the pages of the newspapers with tales of the incompetence of the engineer-in-chief and assertions as to the uselessness of his machinery.

"Through this prolonged storm of abuse," Bennett noted, "Mr. Isherwood passed unconcerned, neither depressed by slander nor exhilarated by success; he had calculated upon both, and was never disappointed."[1] Of all the opposition to Isherwood, the most lively was Dickerson's, so it was natural that attention and support should be drawn to the side of the patent lawyer.

A problem soon arose concerning the proper mode of testing the competing engines. When Forbes had contracted to build the machinery for the *Algonquin,* he had promised to allow the Navy Department to prescribe the test for his vessel. By taking on all the financial risks if the steamer failed to meet the department standards, Forbes had presented such an attractive offer that the department could not have refused him, according to Welles.[2] On this basis, Isherwood determined that the test of the *Algonquin*'s machinery should be based on its ability to generate power, which could best be determined not by a race against the *Winooski* in open water, but by controlled working of the vessels tied to the dock. In this way, the paddle wheels of the steamers would act as "perfect dynamometers," so that the wheels rotating faster would be producing more power.[3]

Isherwood did not consider this test a competition, but rather a trial of Dickerson's machinery to see if the engine, boilers, condenser, and valve gear of the *Algonquin* proved to be better, equal, or worse than the *Winooski*'s machinery. Dickerson, however, had different ideas. Ignoring the contractual provisions, he insisted that the *Algonquin* should only have to achieve a given speed or prove to be more economical than the department vessel. Nevertheless, he was positive that, with equal amounts of fuel, his vessel would develop far greater power. In a public letter which he signed with Forbes's name, he boasted that the *Algonquin* could steam as far and as fast as the *Winooski* at an economical rate of speed and would have enough coal remaining to tow the *Winooski* back to port after she ran out of fuel. To prove the power of his machinery, he would tie the vessels stern to stern, develop a full head of steam in each, and then let them pull against each other, in which case "the *Algonquin* can tow the *Winooski* backward across the North River in twenty minutes."[4]

Irritated with Dickerson's boasts, Welles argued that if the lawyer's engines had merit he wanted to know it. He did not care if they were better than Isherwood's or not; all he was interested in was getting the best engine possible. Dickerson, however, appeared to be all "gasconade and pretension . . . that is flabby and disgusting," as the Secretary recorded in his diary, and he suspected that the lawyer was only concerned with advertising himself.[5]

While New York newspapers criticized Isherwood's dock-trial method of testing the *Algonquin*, Dickerson irresponsibly utilized the name and reputation of his backer to write challenging letters to the Navy Department which Welles treated as impertinent and insolent. One, dated August 7, 1865, was so "improper" that Welles wrote to Forbes, returning the offensive letter while coldly reminding the builder that, according to the provisions of the contract, he had no basis for complaint. By this time Welles had concluded that Forbes was a passive tool of Dickerson and that the lawyer wished to avoid any sort of trial, desiring only to publicize himself at the department's expense. Although Dickerson's letters were designed to provoke retort, Welles wisely decided to stay silent, realizing that the truth would ultimately emerge, despite the fact that Forbes and his "prompter" had control of the New York press and were using it to prejudice the public and to assault the Navy Department.[6]

Forbes, in the meantime, suddenly realized that in aiding Dickerson he had unwittingly opened Pandora's box. Concerned only with the building of his ships, he had ignored the quarrel between Isherwood and Dickerson, but suddenly he had found himself involved. He had taken no part in the attack on Isherwood, he insisted in a letter to Fox, July 11, 1865, and would not do so unless forced. He pleaded with Fox to remove the *Algonquin's* testing from both Isherwood's and Dickerson's hands in order to "see fair play." Only a spectator, Forbes had "no idea of being a victim to their quarrels any more than I can help." When Dickerson began sending insolent letters to Welles, Forbes became even more alarmed and again wrote to Fox explaining that he had nothing to do with the letters. Instead, he insisted, he had been victimized by the crafty patent attorney.[7]

Fox was sympathetic to Forbes and tolerant with Dickerson. He informed Welles that the lawyer undoubtedly was entirely responsible for the letters and that he was writing them just for notoriety, while the papers, in the absence of news, amused themselves by fomenting the issue. Isherwood, however, had different feelings. Usually unmoved by the sniping of his critics, he was disturbed by the letters apparently written by Forbes. He tartly reminded Welles that Forbes, not the department, had solicited the contract for the *Algonquin,* and therefore he had no reason to complain about the stipulations. Forbes's insistence that Isherwood's steam expansion theories were to be challenged in the *Algonquin's* trial bothered the Engineer in Chief even more. He had been fully vindicated in this respect by a congressional investigation, he stated, and there was no reason at all to resuscitate a dead issue.

The accusations which had so annoyed Welles brought an even more heated reaction from Isherwood, since he was the object of the attacks. Forbes's statements about the expressions and intentions of the bureau Chief were unequivocally false, Isherwood declared. They proceeded only from envy and were obviously designed to provide an excuse for the *Algonquin*'s failure. Noting that Forbes had shifted liability for the vessel by subcontracting the machinery with the Providence Steam Engine Works, Isherwood delivered his final stroke in a letter to Welles, dated August 15, 1865. "The Bureau fully appreciates the cheap patriotism which the contractor assumes," he remarked, "and is familiar with its appearance in public business; it being often advanced with a degree of clamor proportional to the want of it."[8]

Sufficiently stirred to repeat his opinion of Forbes in a letter to *The New York Times,* Isherwood went further to accuse Forbes and Dickerson of "villainous calumnies" against him, when his only offense had been "to protect the public treasury against their attempts."[9] By these impassioned accusations, Isherwood was playing into the hands of his critics. *The Times,* strongly sympathetic with

Forbes and Dickerson, immediately replied with two vigorous editorials, declaring that Isherwood's "vulgar assertions" against Forbes revealed that the Engineer in Chief was motivated by "personal hostility" and had exhibited a "general laxity of principle and of self-respect." Isherwood's insulting arrogance was typical of a "subordinate official" who had let his unaccustomed power go to his head when dealing with such an influential and distinguished person as Paul Forbes. The insinuations about Forbes's patriotic motives were particularly reprehensible, *The Times* concluded. Not only did they display bad taste, but they would recoil on Isherwood, because Forbes was too well known and respected for the public to allow such base accusations.[10]

The *Winooski* was ready for the dock trial by the end of August, 1865, but Dickerson still labored over the *Algonquin,* making last-minute alterations. By now he had abandoned any hope that his machinery could develop more power than Isherwood's and relied only on demonstrating more economical operation, the confident Engineer in Chief reported to Welles from New York. Believing that Dickerson, being "very much frightened," would try to avoid the trial if possible, Isherwood suggested that if Forbes were not satisfied with the results of the dock trial he could request a race between the vessels to determine the power developed. In this way, Dickerson might at least go through with the initial dock trials. Without such a duel between the vessels, Isherwood now realized and wrote to the Secretary, "there will be no getting rid of factious people."[11]

In early September, the formal procedure of the dock trial was made public. The two vessels were to run for ninety-six consecutive hours on a given quantity of coal, in order to determine economy of operation. Then they would run for another ninety-six hours on as much coal as they could burn to determine maximum power. Finally, as a concession to the builders, the ships would race the length of Long Island Sound, from Throgg's Neck around Fisher Island and back three times, to determine their relative speed. Supplementing the original examining board of three naval chief engineers, the Navy Department appointed a board of civilian experts. Of the six members of this board, at least two, William Wright and Miers Coryell, were known supporters of Dickerson.

With the trial about to commence, Dickerson became increasingly obstreperous, and Admiral Gregory at the New York Navy Yard warned Fox that the patent lawyer should be brought under control. Gregory was convinced that if anything went wrong with the *Algonquin,* Dickerson would immediately stop the trial, enter a protest, and insist that he was a victim of fraud and injustice.[12]

Back in the Navy Department there was no lack of agreement. Welles told Fox that Dickerson "means to skulk or equivocate, and evade . . . when he began to write and publish his long letters I expected this." More than ever convinced that Dickerson had "most of the New York press in his pay," Welles concluded that this incident was only one of many in which the New York papers were unremittingly belligerent towards the department. As he so often stated, there was a "studied design" behind these attacks, and it was obvious that hostility to the administration as much as faith in Dickerson produced the solid support enjoyed by the patent lawyer.[13]

Isherwood, supervising the preparation of the *Winooski* in New York, believed that the difficulty in getting Dickerson to consent to the trial was because of the remarkable results already demonstrated by the *Winooski.* Among the great number of engineers present, there was no difference of opinion, he jubilantly reported, for all of them admitted the marked superiority of the government machinery. So confident was the Engineer in Chief that he now readily offered to run the test on Dickerson's own terms, allowing him every advantage, even to operating the *Winooski* himself. "She could not be so mismanaged as not to beat him badly even in his own hands," Isherwood smugly concluded in a letter to Fox, September 13, 1865.[14]

Dickerson and Forbes were not ready to surrender. For a week they held up the trial while perfecting their machinery. In an effort to obtain greater combustion in the furnaces they set up airtight bulkheads on either side of the fireroom, and then, in violation of the contract, installed a powerful blower to create forced draft. Unfortunately for Dickerson, the superheated steam required to work his engines at a great degree of expansion produced serious problems, as ten boiler tubes burst under the unusually high pressure. Observing these difficulties, Isherwood concluded that the machinery would be a wreck before the trials were over.

On September 21, the trials finally began. Praying that if Dickerson's "machinery will only hold together a few days longer we shall be rid of it and him together," Isherwood made his appearance early at the Navy yard.[15] Dickerson, "the picture of confidence," attended strictly to his own engine, which had already proved distressingly unpredictable.[16] As the day wore on, the crowd increased to several hundred spectators, drawn to the scene by advance publicity and by the vigorous advertising of local transportation companies. Streetcars running near the dock bore large flags proclaiming, "Great Steamship Engine Trial, Direct!" or, "To the Algonquin and Winooski, Direct!" All day long the excitement mounted, and such was the enthusiasm of some spectators that they had to be cautioned by the press:

> The persons who go on board of these vessels will do the engineers on watch a favor by keeping out of the engine room, as it interferes sadly with their duties, and with the thermometer ranging at about 100, they desire all the fresh air they can get.[17]

By sundown of the first day the crowd had totaled nearly 10,000. Since "betting was freely indulged in," there was such a demand to see the logs of the vessels that guards had to be posted at the entrance of both engine rooms. After the paddle wheels of both vessels had churned side by side at the dock for fifty-one hours, Dickerson suddenly discovered that the feed pipe supplying water to his boilers had failed, so he stopped his engines, claiming that although the *Algonquin* at that moment was trailing in total revolutions, it was gaining steadily on the *Winooski*.[18]

The "accidental" bursting of the *Algonquin*'s feed pipe was greeted with skepticism. A *Times* reporter attributed it to excessive steam pressure and went on to imply that this was not the only reason why both vessels were quite dangerous for spectators. Observing the behavior of the crews with grave misgivings, he remarked that, since the *Winooski* was a government gunboat, at least on *that* vessel the engineers were prohibited by law from drinking liquor "to excess" while on the job.[19] Admiral Gregory dismissed the possibility of accident, insisting that Dickerson had broken down the machinery himself when he realized that his vessel could not beat the *Winooski*.[20]

Although the test of the *Algonquin*'s machinery had not been completed, there was no lack of opinion on which vessel had "won the competition." If some pointed out that the expansion theory should not have been in question and that the determination of power and the cost of producing it was the purpose of the trial, they were ignored. It was far more stimulating to argue whether or not Dickerson had refuted Isherwood's theories, and if there was a shortage of data to substantiate any position, one could always mask it with a bit of choice invective.[21]

John Ericsson did not hesitate to contribute his opinion. Writing to Assistant Secretary Fox on September 30,[22] he listed a number of reasons why he believed that Isherwood had been humiliated by the performance of the *Algonquin*. Admitting that Dickerson was incompetent as a design engineer, Ericsson attempted to prove that despite the eccentric ideas of the patent lawyer, his basic theory was correct because his machinery had performed so well during the time it was able to run. To emphasize this, Ericsson had to argue that the *Winooski* had outperformed the *Algonquin* only because Isherwood had used a superlative engine which produced such great economy and efficiency that no vessel afloat had better or more economical motive power.

Handing this letter over to the Engineer in Chief, Fox sat back to watch the result; Isherwood did not disappoint him. On October 6 he sent Fox a fifteen-page letter which, point by point, answered the allegations of Ericsson.

In the first place, Isherwood noted, the test had been unfair to him, since he had not been told it would be one to measure economy until after the *Winooski*'s engines were under construction, so Dickerson had an advantage in employing a specially designed engine against one which had been built for durability and reliability, rather than economy. Dickerson used superheated steam at seventy-one pounds pressure in a small engine which he could work to capacity in order to produce his most economical results. Isherwood, on the other hand, had a heavy, large cylinder engine which used low-pressure, saturated steam at small measures of expansion.

The *Winooski,* designed to produce 1,500 horsepower, had to operate at only ⅓ power in order to match the *Algonquin*'s maximum of 500 horsepower. This was obviously disadvantageous for the *Winooski*, since Isherwood had used steam at only 17 pounds pressure to keep down the power developed. It was foolish for Dickerson to insist upon a race, the Engineer in Chief reminded Fox, since his engines, which weighed as much as Isherwood's, produced only

⅓ of the power, so the *Algonquin* at best could go only 9½ knots against the 13¾ knots the *Winooski* had already demonstrated in trials.

In attempting to argue that Isherwood's steam expansion theory had been exploded, Ericsson had trapped himself, the Engineer in Chief concluded. If the *Winooski* had the best engine in the world, as Ericsson had claimed, how could Isherwood's theory be fallacious? After all, he had designed the engine on his theories, and it was not even special machinery. Ericsson was making a fool of himself by trying to twist the facts to suit his own fallacious theory. Dickerson, Isherwood wrote to Fox, faced with his failure to demonstrate greater fuel economy, equal power, or equal durability and reliability, had been wise to break his engines down "to save himself from ignominious defeat."[23]

Dickerson, however, had not yet finished. After extensive repairs on his engines, he once more brought the *Algonquin* against the *Winooski* in an attempt to shatter the Engineer in Chief's confidence. On October 23, the two vessels started off on their second ninety-six-hour trial. This time the *Algonquin*'s engines fared somewhat better, not stopping until almost seventy hours had elapsed. Immediately a storm of dispute rose over the outcome of this trial. In this second test, Dickerson had stopped his engines for no apparent reason, other than his dissatisfaction with the progress of the trial, but he subsequently discovered an ingenious explanation for his actions. During the course of the trial, a loose coal barge had accidentally drifted into the *Winooski* and slightly damaged one of her paddle wheels. This, Dickerson insisted, subsequently lowered the resistance of the water to the wheel, and gave the *Winooski* an unfair advantage over his vessel. Unimpressed with this argument, the board of naval engineers reported to Welles that once more the *Winooski* had demonstrated her superior machinery, aside from actually completing the test, an achievement still beyond the capacity of the *Algonquin*.

Paul Forbes now insisted that a dock trial was quite unfair and improper as a test, since his engine, he maintained, "was built to navigate the sea, and not to strain hawsers and pull docks." It was a sea trial that he wanted, and whenever the Navy Department would oblige him, he would have the *Algonquin* ready.[24]

To satisfy the contractor once and for all, the department finally succumbed to public pressure and agreed to make a final comparative test of the *Algonquin* and *Winooski* on Long Island Sound, racing them 8 times consecutively over a course for a total of 904 miles. The date of the race was delayed until February, 1866, since Dickerson had to repair the *Algonquin*'s machinery from the ravages of the dock trials.

By February 1, Isherwood was in New York, impatiently waiting for Dickerson to complete his preparations. Contemptuously, the Engineer in Chief reported to Fox that Dickerson and Forbes were having difficulties trimming their vessel, since Dickerson had placed most of the machinery on one side, resulting in "an enormous list to starboard." By February 11, Isherwood began to doubt if the *Algonquin* would ever be ready, observing that Forbes had just withdrawn all his engineers from the vessel. Isherwood, on board the waiting *Winooski,* told Fox that the only way to end such trifling was to run the *Algonquin* by using the naval engineers who were on board as observers.

The Engineer in Chief's pessimism proved to be short-lived. Within two days, Forbes and Dickerson pronounced their vessel ready to race, and at 3 P.M. on February 13, the two vessels moved into the chilly, ice-strewn waters of Long Island Sound.

While some of Dickerson's advocates, perhaps anticipating more trouble with the *Algonquin,* objected to this trial of speed as irrelevant to contractual stipulations, most observers agreed with the *Scientific American,* of February 17, 1866, that "if the *Winooski* beats the *Algonquin,* Isherwood's theories are correct. If the *Algonquin* beats the *Winooski,* Dickerson's theories are correct. Engineers will govern themselves accordingly...."[25]

The race began in smooth water with only a gentle, variable breeze blowing, but by the evening of the following day a "violent storm" arose, buffeting both vessels with high winds, rain squalls, and spray which spread a coating of ice over all exposed surfaces. Floating ice at both ends of the course further impeded the progress of the vessels by greatly reducing their speed when turning.[26]

Quickly the superiority in power of the *Winooski*'s machinery became glaringly evident. Forty miles from the starting point the government vessel was already four and one-half miles ahead of her rival. Gaining steadily on the *Algonquin,* she was thirty miles ahead by seven o'clock the following morning, and by the evening, when she had completed her third round trip, she was leading by seventy miles. At this point, weather conditions had become so severe that the race was halted, and the *Winooski,* with Benjamin Isherwood aboard, triumphantly steamed back to port, a convincing victor.

The machinery of both vessels was in excellent order at the finish of the race. The *Winooski* had functioned flawlessly with "regularity, noiselessness and smoothness of motion." The *Algonquin* had suffered minor breakdowns, including a broken paddle on her way back to port, but nothing had impeded her performance. Although Dickerson had used blowers for forced draft, where the *Winooski* had none, his engines had failed him. By placing them so poorly in the hull, the patent lawyer had been forced to add 73 tons of ballast to correct a 22-degree list, and even before this addition his machinery had already outweighed Isherwood's by almost 45 tons.

Dickerson's vessel averaged 9.869 knots to the *Winooski*'s 11.737 knots, although carrying nearly twice as much steam pressure in her boilers. In efficiency, the disparity was just as large—the *Algonquin*'s machinery, with a .126 cutoff, using 4.122 pounds of anthracite coal per horsepower hour while the *Winooski* used 3.476 pounds of coal, with a .705 steam cutoff. The *Algonquin* could develop only 54.29 per cent of the *Winooski*'s power despite being "pushed to her utmost." Isherwood claimed that his machinery was not extended and that the vessel could have made 1 to 1.5 knots better speed with blowers and higher steam pressure. He assured Fox, in a letter of February 16, that if the trial had continued to its conclusion, "we should probably have towed back the *Algonquin* to port."[27]

The board of naval engineers agreed with Isherwood. On every point guaranteed for the *Algonquin*'s machinery, Dickerson and Forbes had failed. Their machinery was much inferior in materials and in workmanship, was 39 per cent heavier with the ballast, took up 11 per cent more space, and thus correspondingly decreased the room for coal. The *Algonquin,* the board concluded, was "totally unfit for the naval service."[28] Welles demanded that the machinery be removed and replaced with the same type as that in the *Winooski,* and he withheld the final payment to Forbes on the contract, an act which prompted the latter to initiate an appeal to Congress for leniency which proved successful only after Welles had departed as Secretary.[29]

The results of the *Algonquin-Winooski* race largely disillusioned

Dickerson's supporters. In England, where engineering journals reluctantly had encouraged the patent lawyer because of their distaste for Isherwood's engineering theories, the outcome of the race provoked extreme disgust. American engineering, they declared, was hopelessly backward in permitting a dispute over steam expansion, a principle long settled in English circles. The tragedy was that the man arguing for the proper theory was an utter fool, and his defeat thus assured the perpetuation of ignorance among American engineers. "Here in England," the *Engineering* journal announced, "some of Mr. Dickerson's *quasi* professional effusions would be held abominable, and it would be difficult for their author even to maintain a position in good society."[30]

Dickerson, however, was not yet silenced. Brushing off the debacle of the *Algonquin* as inconsequential, he once more challenged the Engineer in Chief with the oncoming trial of the cruiser *Idaho*. This vessel by contract had to make fifteen knots for twenty-four consecutive hours in order to be accepted by the Navy Department. As with Dickerson's other vessels, the *Idaho*'s machinery was designed to utilize steam at very high measures of expansion, and also as with his other vessels, her machinery was of a bizarre design.[31]

On May 15, 1866, the *Idaho* went to sea with the stated purpose of breaking existing speed records. As usual, Dickerson's machinery at once broke down, and it was not until May 19 that the 24-hour test began in a light wind and moderate swell. At the conclusion of that day's run, the ship had achieved an average speed of 8.27 knots, and the board of naval observers concluded that the *Idaho* was not an efficient vessel of war and that "her machinery is worthless as a motive-power." Unwilling to retain her in the Navy, they recommended that the ship be sold at auction, estimating that the proceeds would be $125,000. This would result in a considerable loss for the department, which had already expended $550,000 in progress payments on Forbes's cruiser.[32]

Isherwood was not surprised. He had examined the ship in April and doubted if she could do better than eight knots, if, indeed, she could be kept in one piece for that long.[33] "In fact," he remarked in a letter to Fox, dated April 24, 1866, "a more complete failure in every particular could not be made. I did not think it possible, at this era, to build marine machinery so totally worthless. Every dollar spent on it is so much thrown away."[34]

Although Forbes had solicited the contract for the *Idaho,* had made his own terms, and had forced the Secretary into accepting the offer, he refused to accept defeat and the loss of money it entailed. While Dickerson disappeared from public view, Forbes kept up a steady political pressure on Congress until it finally voted, on February 18, 1867, to accept the vessel for the sum already paid to Forbes, only $50,000 less than the original contract price. Ironically, after the Navy Department removed Dickerson's machinery and turned the vessel into a full-rigged sailing ship, she performed brilliantly, logging speeds which would establish her as one of the fastest sailing ships in the world.[35]

Dickerson, the eccentric and imaginative monomaniac, ceased to challenge and abuse the department and its Engineer in Chief, and for this Isherwood could breathe a long sigh of relief. Advances in technology were soon to make high degrees of steam expansion necessary to engine design, but Isherwood, nevertheless, had still made a significant contribution to his field. By publishing his experiments and then proving their validity in competition with a man who embodied the weaknesses of unscientific design, Isherwood enabled his profession to advance on a sure and practical footing, free from the excesses and fallacies of irresponsible theorists.

# VII. The Steam Bureau under Attack

During the Civil War the financial transactions of the Bureau of Steam Engineering became a matter of intense public interest. To achieve wartime goals in the construction and repair of steam machinery, the bureau necessarily spent tremendous sums. This outlay of government funds attracted hordes of private builders and suppliers to the Navy Department.

In his 1862 report to Welles, Isherwood estimated that for the fiscal year ending June 30, 1864, his bureau would need $5,800,000, of the $68,257,255 of congressional appropriations called for by the entire Navy Department. At this time the steam bureau stood a modest fourth in order of the amount of funds requested, far behind John Lenthall's Bureau of Construction and Repair, which had asked for $24,598,000.

Within a year there was a dramatic change in position. For the fiscal year ending June 30, 1865, the Navy Department requested $142,618,785, slightly more than double the amount of the previous year. Isherwood, however, asked for $39,362,000, almost $7,000,000 more than Lenthall required for his bureau's operations. The Bureau of Steam Engineering, with nearly a seven-fold increase, now stood first in order of naval requisitions.[1]

With the sudden expansion in machinery building, private contractors were understandably annoyed when the Chief of the Steam Bureau insisted on designing so many of the naval engines himself, stipulated Martin boilers and Sewell condensers to the exclusion of other models, and drove a hard bargain with any private builder or supplier who held a contract with his bureau. By 1863, in an attempt to pry loose from his control the design and construction of naval machinery, they had launched a vigorous attack on Isherwood.

Led by Edward Dickerson, private contractors brought pressure on the press and on Congress in an attempt to challenge the Engineer in Chief. The engineering knowledge of the world, they maintained, was not centered in his bureau, and they censured the professional exclusiveness and conservatism there which had produced "our slow and rolling official 'tubs' upon the advent of the Alabama." Isherwood, his critics insisted, could not build fast ships; and without speed, naval blockading vessels were worthless in dealing with the Confederate blockade-runners.[2]

Such pressure was brought on Welles that he agreed, in January, 1863, to establish a board of civilians who would examine Isherwood's steam machinery. The Secretary ordered Isherwood to attend the meetings of this board, to submit to them the plans of all machinery proposed and built during the previous two years, and then to be on call whenever a question arose.[3] As this board included civilian engineers, some of whom were known to oppose the bureau Chief, it was not surprising that the report which they submitted in March criticized Isherwood's work.

His machinery was too big for the amount of power it developed, they decided, and Martin's boiler was inferior to comparable models. In addition, they were of the opinion that Isherwood and Lenthall's plan for the 7,500-ton ironclad cruiser was quite impracticable. Isherwood, however, made the best of a bad situation. Since the board had been asked to examine boilers, he joined the experts, insisted on further boiler experiments, and kept the group at work until the end of 1868 performing elaborate tests and contributing to his work on steam expansion.[4]

At the same time, Senator Orville H. Browning, impressed by Dickerson's campaign, introduced a congressional resolution intended to embarrass Isherwood by investigating the *Pensacola* affair and the speed of the *Ossipee*-class sloops. Angered by this harassment, Isherwood took enough time from his duties to compile the checkered history of the *Pensacola,* and also to demonstrate that the *Ossipee* class had sufficient speed, since one ship had made eleven and two-thirds knots without blowers, and another had achieved twelve and one-half knots on her trial trip.

Such attacks were not all directed solely against Isherwood, for even when he was the victim, he frequently had not been the prime target. There never had been a unified support of the Lincoln administration in the North; and as the Civil War progressed, opposition grew bitter and more effective. Any segment of the government might come under attack if it were engaged in controversial operations, and the Navy Department was particularly vulnerable. Moreover, it was not only anti-Union cirticism which plagued the Navy. With struggles for power within the President's cabinet added to jealousy and hostility in Congress, Gideon Welles often found his department to be the focus of abuse from many directions.

By February, 1864, Welles had decided that the "persistent assault" on the Navy Department was far from unorganized. Although he could produce no evidence, his long exposure to political infighting convinced him that the War Department, led by Secretary Stanton, had been "secretly instigating" attacks on the Navy Department in order to cover up their own shortcomings. Welles had disliked Stanton from the start, and their disagreements over joint Army-Navy operations, particularly in the West, had exacerbated this hostility. Welles thus concluded that Stanton and his minions were behind the continuing attacks of contractors, claim agents, unprincipled congressmen, and the corrupt members of the press.[5]

In Congress, matters were no better. Senator John P. Hale, of New Hampshire, the chairman of the Senate Naval Affairs Committee, had proved to be particularly trying. Instead of supporting the Navy Department by promoting its interests in Congress, he had become its bitter opponent, apparently after failing to obtain naval contracts for his friends. While several congressional investigating committees pored over documents and testimony seeking evidence of corruption and incompetence in the Navy Department, Hale, as a perverted Diogenes, led their efforts, "searching, as with a lantern, for errors and mistakes." Infuriated by Hale's harassment, Welles accused him of employing "detectives, rotten and disappointed contractors, and grouty party men of the Navy, as well as politicians of every kind of politics to aid him. . . ."[6]

Between February and June, 1864, Hale served as chairman of a committee which investigated the possibility of corruption, irregularity, and favoritism in contracting for naval supplies. Spurred on by pamphlet attacks on Isherwood's and Lenthall's bureaus, the committee examined the bidding and negotiation procedures; but despite Hale's hostility, little evidence was produced to implicate the Navy Department itself, and Isherwood and Lenthall were cleared of any imputation of corruption.[7]

While bearing up patiently under congressional investigation, Welles also had to deal with the press. The center of attack on the Navy Department was New York city. Antiwar sentiment, hostility toward the Republican administration, and an insatiable demand for sensationalism combined to provide a steady flow of diatribe which was successful in creating public resentment toward the Navy Department. Once again scenting a "covert and systematic attack on the Navy Department," Welles showed little partiality in condemning the New York papers. In his view, Horace Greeley of the *Tribune* was secretly hostile to Lincoln; the publisher of the *Evening Post* was under bail for embezzlement and fraud which the Navy had refused to conceal; and *The Times* was merely an organ for Seward and Thurlow Weed. With its editor, Henry Raymond, an "unscrupulous soldier of fortune," Welles felt that *The Times* was the most biased of all against the Navy Department.[8] As the elections of 1864 approached, Welles grew increasingly short tempered, finally concluding that "there is not an honest fair-dealing Administration journal in New York."[9]

In congressional debates, Isherwood was rarely spared as the Navy Department came under frequent attack. Next to Senator Hale, the most vigorous opponent was Representative Henry Winter Davis, of Maryland, who ridiculed Isherwood's incessant experimentation, called Fox the real and acting Secretary of the Navy, and asked for a Board of Admiralty to rescue the decision-making from the triumvirate who were presently ruining the Navy—Welles, Fox, and, Isherwood.[10]

Isherwood and the Navy Department, however, had a few defenders. In the Senate, James W. Grimes, who would succeed Hale in 1865 as chairman of the Naval Affairs Committee, was one man on whom Welles could usually depend for aid. On February 1, 1864 he vindicated the Secretary's trust. Grimes read to the Senate a long letter written to him by the distinguished clipper-ship designer and builder Donald McKay, and in this letter Benjamin Isherwood at long last received public credit for his efforts.

Barely acquainted with Isherwood and having been considered previously as very unfriendly towards the Navy Department, McKay believed that his own words should carry considerable weight. American naval engineers, he declared, were "an ornament to their country" and were better than their English counterparts. Isherwood was a man of great practical experience and thorough education, and as most engineers and manufacturers would willingly admit, a "sound, clear-headed thinker." His books had been extensively read abroad, and his experiments had won from foreign experts their "very high opinion of his capacity as a marine engineer."

Absorbed with his work, Isherwood had been unfairly bothered by monomaniacs like Dickerson and by idlers who had little else to do, argued McKay. Isherwood's engines, he continued, had been properly designed for their purpose, for the Engineer in Chief's emphasis on durability and simplicity was not only necessary for wartime conditions, but was also a valid construction principle for any period, as proven by the success of English machinery built in the same way. Private American builders and designers had not equalled Isherwood's results even when their efforts had been responsible, so an attack on the Engineer in Chief could not be justified on grounds of his incompetence. The irresponsible critics should be the ones to receive punishment, McKay urged, since all they had accomplished was to blacken the department's reputation and please the English.

With a judgment based on his own success as a designer, McKay guaranteed to Grimes that the new class of supercruisers being built for the Navy would have no equal for speed and seaworthiness. He urged construction of fifty such vessels with Isherwood's machinery in every one of them.[11]

Before McKay's letter appeared, Henry Winter Davis, at the instigation of Dickerson, presented a resolution before the House of Representatives on January 7, 1864, demanding a full investigation of Isherwood's engines. Despite the obvious intent to embarrass the Engineer in Chief, the House approved this resolution, since the controversy over Isherwood's theories and designs had become so bitter and so widespread that Congress apparently decided to settle the issue by a thorough examination of all the work he had performed since coming into office in March, 1861.[12]

Davis's resolution, showing the patent lawyer's influence, was extremely hostile in tone. It demanded information about the degree to which Isherwood's engines differed not only from those used in commercial steamers, but also from those in English and French warships, and "whether their inadequate power and speed are caused by such differences." Davis also intended to determine if unfair practices had been employed in the *Pensacola*'s trial "with a view to break it down and bring it and the plan on which it was constructed into disrepute." Finally, he intimated that someone in the steam bureau was receiving pay-offs from engine contractors, and was forcing the payment of patent fees to those not entitled to them. To make sure that there would be no confusion about who was the object of these accusations, Dickerson appeared before the investigating committee and read to them his *Navy of the United States*.[13]

The House Committee on Naval Affairs, with Alexander H. Rice as chairman, spent nearly a year examining more than forty witnesses and collecting data on all the ships powered by Isherwood's machinery. On January 31, 1865 they completed their work and presented to Congress the report which was to serve as an official evaluation of Isherwood's career as engineer in chief during the Civil War.

In the first place, stated Chairman Rice, the resolution which produced this investigation had made no direct charge against Isherwood, but had merely implied his failure, incapacity, and fraud. Moreover, the public hostility towards the Engineer in Chief was vague and general in character, rather than being a definite opinion based on any "palpable failure" of the Navy, either in whole or in part. For this reason, the committee had to cover a larger area than they had originally intended, since there were few specific items or accusations with which they could deal directly.[14]

A chronic complaint against Isherwood had been that his vessels were too slow to be of any use as warships. The committee ran only one speed trial on an Isherwood vessel, the double-ender *Eutaw*, and discovered that she developed a maximum speed of 13.7 knots. Using superheated steam, she could average 13.1 knots consistently, while consuming coal at the very economical rate of 2.323 pounds per horsepower hour. Comparing the performance data of Isherwood-powered naval steamers with that of other ships, the committee concluded that the speed of American naval steamers had not decreased since Isherwood came into office, but had actually increased to a considerable extent. Prewar steamers, they found, had maximum speeds of 8 to 11.7 knots under steam alone, and had averaged 6.5 to 8.5 knots under both steam and sail. The fastest ocean-going merchant steamers averaged 10.5 knots on a continuous voyage, although the famed American Collins liners could run at 11.75 knots, using both sails and steam.

By comparison, Isherwood's machinery stood up well. The *Ti-*

*conderoga* class of screw sloops had a maximum speed of twelve knots, the *Nipsic* class of screw gunboats did twelve and one-half knots, while the *Itasca* class made ten knots. Of the double-enders, the earlier class could steam at a maximum of eleven knots, while the later vessels, such as the *Eutaw*, were capable of more than thirteen and one-half knots.

The general complaint that Isherwood had monopolized the design of naval machinery had no valid foundation, the committee decided. Before the war there were eight different engine designs employed in the twenty-three ships built for the Navy, while during Isherwood's period there were eight different designs among thirty-seven screw vessels constructed, although all of the thirty-nine side-wheelers were powered by his engines. One had to remember, the committee pointed out, that in wartime some standardization of design was necessary, since time did not permit leisurely experimentation. If they worked well, engines had to be used even if they were not ideal. Considering the conditions of war, the committee concluded, a relatively large number of engineers had contributed to machinery design.

The engines designed by the Engineer in Chief made no radical departure from previous models. Reviewing the testimony of a number of naval engineers and builders, the committee decided that Isherwood should not be accused of irresponsible innovation. Similarly, the Martin boiler was not new, having been the standard for the American Navy before the war, and private builders still preferred to use it unless they had a design of their own. Also, the committee noted, the Navy Department had been willing to try out and occasionally to use other boiler designs for their naval vessels.

The real engineering controversy was not over the Martin water-tube boiler versus the fire-tube boiler, but rather over the kind and size of cylinder used in marine engines. Isherwood used small cylinders with relatively large boilers, since he believed in only moderate use of expanded steam. As a result, the steam cutoff and the use of steam expansively in the Navy had changed little from prewar practices. Although Dickerson, on the basis of Mariotte's law, hotly disputed this policy, he did not receive any endorsement or encouragement from the committee. Relying on the professional opinion of the National Academy of Science, the committee generally approved of Isherwood's findings, stating, "the evidence . . . showed that the theory of the Bureau of Steam Engineering is supported by a very great preponderance in actual practice. . . ."[15]

Criticism of the weight of Isherwood's machinery was also unfounded. Heavier engines, as Isherwood had often insisted, were necessary for durability, reliability, reduction of friction, steadiness, and continuous operation at maximum power. Frequently, the committee added, the critics' own engines proved to be even heavier.

Justifying the use of the Sewell condenser and the valve gear employed by Isherwood, the committee decided that there was no evidence of corruption or graft involved in the choice of these items. Exhaustive hearings had failed to produce proof that any patent payments had been improper, and no witness could present a single example to support the assertion that contractors had been forced to bribe the Engineer in Chief in order to get their machinery used by the Navy. No fees or commissions had been illegally received by the steam bureau, and Isherwood had not practiced extortion.

On the basis of these results, the committee concluded that American naval engines were as good as British or French machinery and, in fact, were very similar to the former in design. Isherwood, far from being a radical innovator, was conservative in both design and practice; and it was such critics as Dickerson, the committee observed, who brought forth the radical theories and engines. Isherwood's engines, instead of lacking in power and speed, showed a "great increase," because of his improvements over former designs. In light of these results, the committee was highly dubious of any designs exalted over those of the Engineer in Chief.[16]

Even a cursory reading of the resolution initiating this investigation revealed that it was in the nature of a charge and an accusation against Isherwood, the committee noted. The resolution had implied that the Engineer in Chief had made radical and detrimental changes to marine engines, to the great misfortune of the American Navy, and that his incompetence and irresponsibility had been accompanied by a corruption deserving of censure. This, the committee declared, was wholly unfounded, there being not a shred of evidence to support such accusations. Instead, the mass of testimony indicated "conclusively that in enterprise, desire for improvement, and devotion to the public service, the Navy Department has not been exceeded by any other branch of the government since the breaking out of the rebellion."[17]

No matter how conclusive and authoritative, this congressional report did not put an end to the criticism of Isherwood's engineering designs and theories. The defeat of the *Algonquin* would finally silence Dickerson, but there still existed a far more formidable opponent than the patent lawyer. In the postwar years, Isherwood would face a challenge by a man of equal skill, equal resourcefulness, and equal persistence. In John Ericsson, Isherwood met his greatest engineering rival.

# VIII. The Inventor and the Engineer

In many ways, John Ericsson closely resembled Benjamin Isherwood. Like the Engineer in Chief, the Swedish-born inventor was a strong, heavy man with a commanding, uncompromising personality. He also combined a fertile, inquiring mind with an energy that could translate dreams into reality. Not only had Ericsson the ability to inspire in his associates an unshakable conviction of his genius, but like Isherwood, he could fire men's minds and energies to a peak of achievement. For years he, too, had suffered abuse and ridicule from the press and the public, although after 1861, Ericsson became the toast of his time as the father of the *Monitor*.

Unlike Isherwood, Ericsson was primarily an inventor. His preoccupation was with design, not construction, and once the design was complete he often lost interest in his project. After the plans left his hands and went to the builder, he showed little curiosity in the details of construction, and he was similarly indifferent to the later practical operation of his creations. It was this attitude which often compromised the quality of his work and betrayed his best efforts. Disregarding the practical value in the operation of machinery, he frequently made errors which could have been avoided. His drawing board, it has been observed, was not the best place to solve his engineering problems.

With his indifference to the practical workings of his inventions, Ericsson combined an intransigency which blinded him to the possibility of error. Unlike Alban Stimers or Edward Dickerson, Ericsson was truly a great engineer and inventor, but with those two he shared a fatal inability to recognize or admit his own mistakes.

While Ericsson enjoyed the encouragement and support of many influential men who were critical of Isherwood, there was one individual in particular whose admiration for the inventor produced trouble for the Engineer in Chief. This was the militant journalist William Conant Church. In August, 1863, Church published the first issue of the *Army and Navy Journal,* a weekly newspaper which served for decades as the leading spokesman for the American military services. Throughout the Civil War and for years afterwards, the *Journal* flourished as a semiofficial organ, and in its pages the controversial military and naval issues of the day received a thorough, if often biased hearing. As editor, Church wielded enormous power, and he was destined to do both good and evil with his vigorous and often biting judgements.

As his biographer has remarked, Church "consistently supported the reform elements within the army, but did little except hinder the creation of the new navy."[1] Not a professional military man, Church leaned heavily on the advice of certain individuals whom he chose as his experts. Often depending on naval line officers to formulate his editorial policy concerning the Navy Department, Church unfortunately made a bad choice in his advisors. He threw in his lot with the reactionaries, and this initial alignment with the naval "standpatters" was a blunder which Church was never able to erase.

While he chose to follow the advice of those line officers who opposed naval progress throughout the later part of the nineteenth century, he also became captivated by the accomplishments of John Ericsson, and it was to Ericsson that he consistently turned for technical advice about the construction of naval machinery and vessels. Consequently, Church was an ardent supporter of the monitor design and of the passive coast-defense strategy based on this type of vessel, and he was implacably hostile to those who opposed the monitor or its creator.[2]

Benjamin Isherwood, therefore, drew the attack of Church on several accounts. In charge of naval engineering, he frustrated the attempts of civilian inventors and builders, especially Ericsson, to provide vessels of their own designs for the Navy. Isherwood's seagoing casemated ironclad plans were in direct opposition to Ericsson's monitor program, and the two had disagreed fundamentally on the question of steam expansion. For Church, this was more than sufficient reason for action. Aided by Ericsson, and on the basis of his own convictions, Church began a steady persecution of Isherwood which lasted throughout the Engineer in Chief's naval career.

The antagonism between Ericsson and Isherwood, however, developed slowly. Despite the fact that Isherwood initially rejected his monitor concept and then, when using it, had the bad grace to employ an Englishman's turret design, Ericsson did not attempt to oppose Isherwood's nomination as bureau chief in 1862. As the inventor mentioned to his friend W. L. Barnes, it mattered not who was chief of the Bureau of Steam Engineering as far as his contracts with the Navy Department were concerned. Ericsson would soon lose this attitude of indifference.

His main divergence of opinion with the Engineer in Chief in 1862 was over engineering theory. Ericsson rejected the conclusions of Isherwood's Erie experiments and declared that Isherwood had produced "an enormous waste of motive force," by insisting on large boilers and small cylinders for naval machinery. Concluding that Isherwood's theory of expansive steam was a "monstrous fallacy," Ericsson became outraged when the Engineer in Chief criticized his monitor engines in *Experimental Researches*. Isherwood's contention that condensation in the cylinders resulted in a large power loss was absurd, Ericsson declared, and he warned Fox that he would not quietly submit to such irresponsible statements.[3]

When the Navy Department announced plans for its class of supercruisers in 1863, Ericsson quickly wrote to Fox, informing the Assistant Secretary that he was "very anxious" to try out an engine of his own design against the geared engines Isherwood planned for some of the vessels.[4] Ericsson's insistence on building a competing engine, added to similar requests from other builders, came at an appropriate time. The Navy Department, already sensitive to accusations of favoritism toward the Engineer in Chief, decided that it would be injudicious to have the entire class of cruisers powered by Isherwood's machinery. Consequently, the department permitted the construction of three cruisers by private builders; Forbes's *Idaho*, which had a hull designed by Henry Steers and used Dickerson's engines; the *Chattanooga*, with a hull built by Cramp and Sons and machinery designed by Merrick and Sons; and the *Madawaska*, which had a hull designed by B. F. Delano and had engines by Ericsson.

Of the three privately built vessels, only one, the *Madawaska*, was to have a hull identical to those of the three Navy Department cruisers, the *Wampanoag*, the *Ammonoosuc*, and the *Neshimany*.[5] From the beginning, Ericsson had demanded an identical hull for his engines. In this way, the inventor reasoned, there could be no excuse for Isherwood's failure when the *Madawaska* proved to have greater speed and economy than the department's vessels.

Ericsson soon found cause for complaint about the contractual arrangements. In his desire to insure a direct competition between his engines and Isherwood's, the inventor had agreed to design only an engine and propeller, the rest of the machinery being the same as that used by the Navy Department. However, this would force him to use boilers of an "enormous size," in accord with Isherwood's practice. This was ridiculous, Ericsson maintained, since the boilers would supply far too much steam for the size of his engines, which, like Isherwood's, were to have cylinders measuring one hundred inches in diameter with a four-foot piston stroke.[6]

The inventor had the same difficulty in other vessels he hoped to build for the Navy Department. As much as he wished to compete with the Engineer in Chief, he felt frustrated by Isherwood's insistence on large boiler power, regardless of engine size. With a limited space for machinery in any vessel, Ericsson explained to Fox, there was insufficient room for large engines, since Isherwood's boilers had to come first.

While competing in machinery design with the Engineer in Chief, Ericsson put his major efforts into the various monitor vessels he was building for the Navy Department. Even with these he was not free from the control of the Bureau of Steam Engineering. Although it would be incorrect to say that Isherwood was persecuting the inventor, the continual modifications in design and the rigid standards in construction which Isherwood insisted upon drove Ericsson to distraction and convinced him that the Engineer in Chief had singled him out for abuse.

This situation was perhaps inevitable. Ericsson, interested only in his basic designs and proud of his patriotic efforts, refused to tolerate the continual restrictions and corrections placed upon his work. Isherwood, from the stream of complaints and suggestions flowing in from the fleet, was concerned solely with assuring maximum utility for the monitor vessels, and he had no patience for anyone who refused to agree to modifications resulting from the fleet's experience. The two men continually clashed, neither being sympathetic with the other's predicament and both unwilling to take the time or trouble to sit down and reach a *modus vivendi* with his professional rival.

With this background of disagreement, and with the *Army and Navy Journal* zealously adding fuel to the fire, the *Madawaska*'s

*Secretary of the Navy Gideon Welles.*

THE NATIONAL ARCHIVES

*John Lenthall, chief of the Bureau of Construction, Equipment, and Repairs.*

THE NATIONAL ARCHIVES

*Edward Nicoll Dickerson.*

THE NEW YORK PUBLIC LIBRARY

*John Ericsson.*

UNITED STATES NAVY

*Gustavus Vasa Fox, assistant secretary of the Navy.*

THE NATIONAL ARCHIVES

*The Sloop-of-War* Wampanoag, *the fastest ship in the world. From a steel engraving in* Harper's Weekly, *June 9, 1866.*

HARPER'S WEEKLY

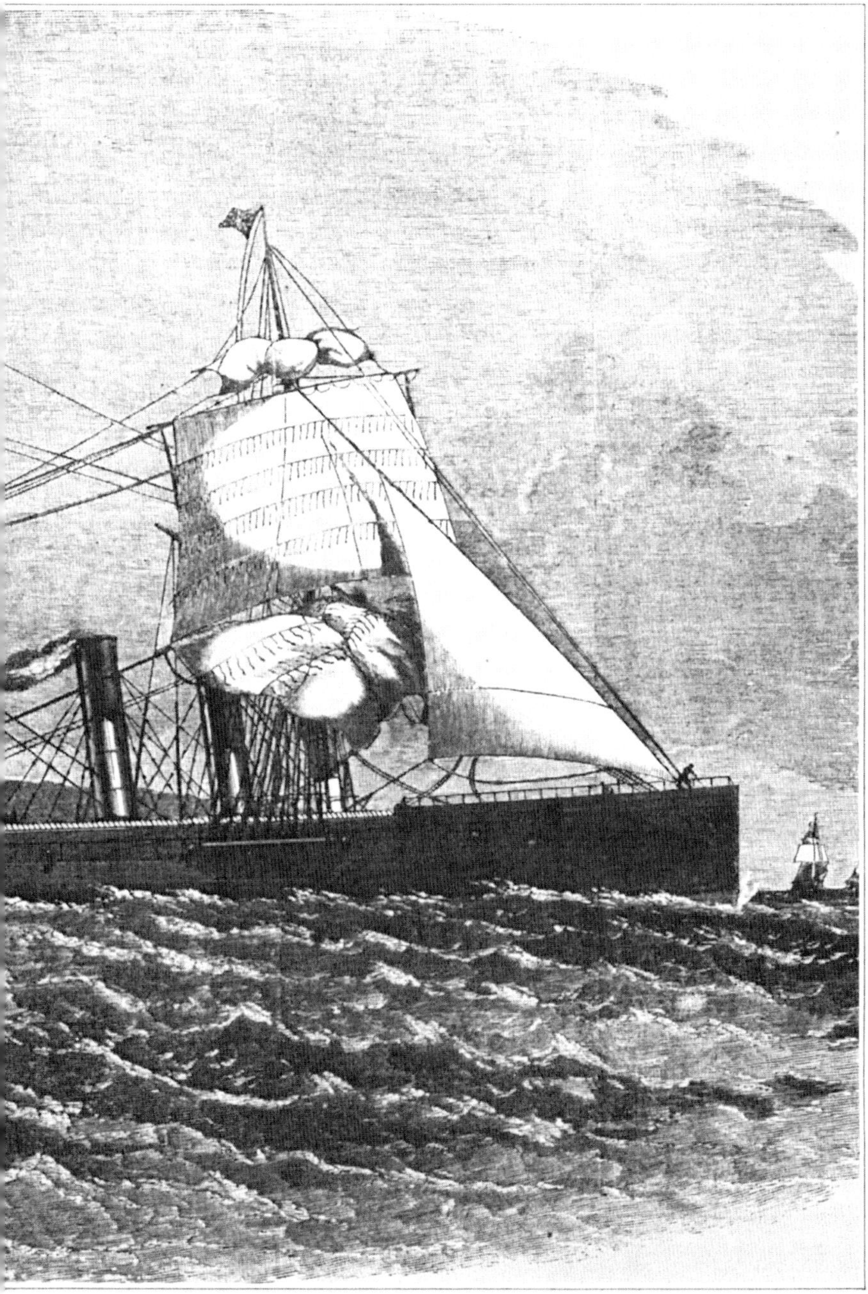

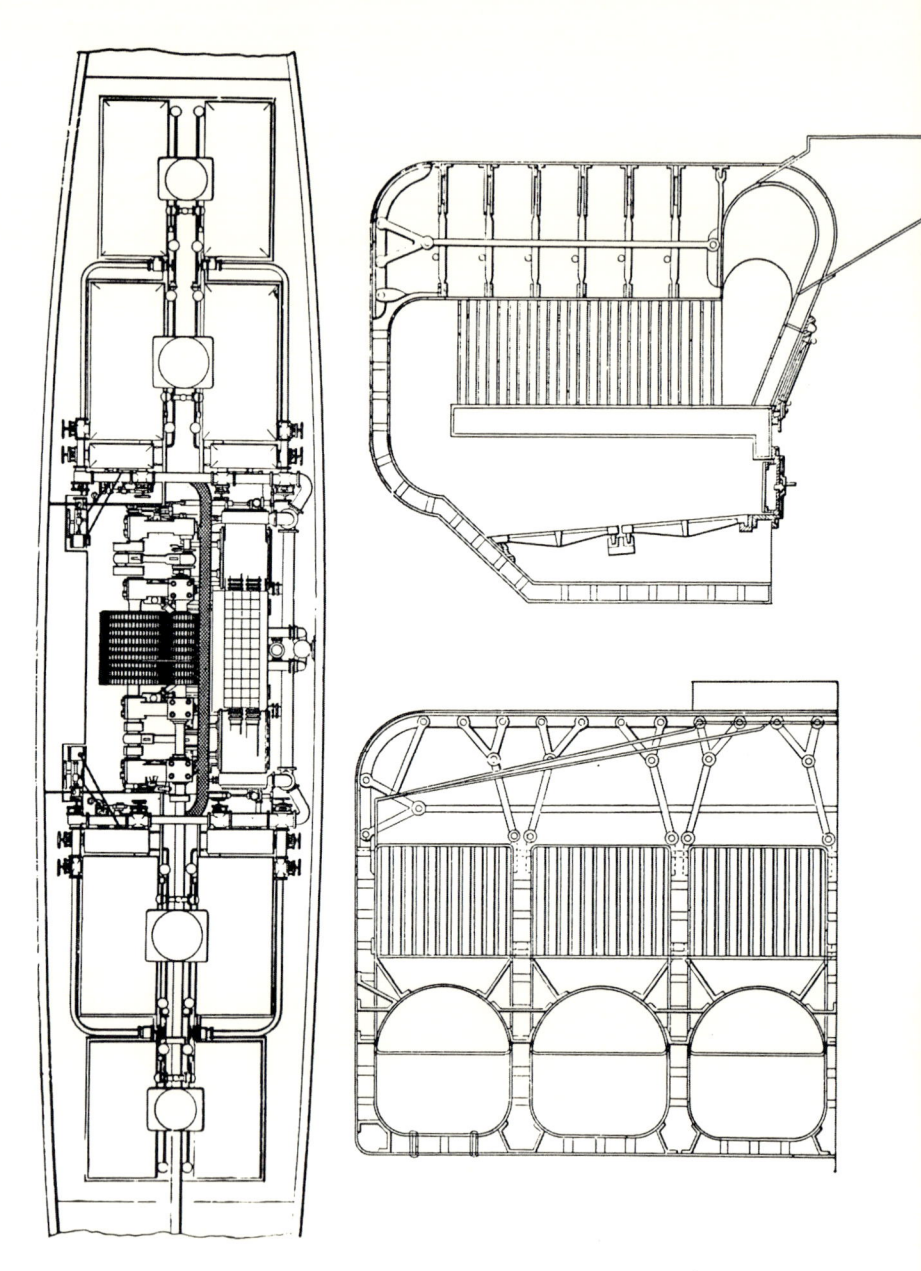

*Examples of Isherwood's concern with detailed specifications for engine design, as taken from his article "The Sloop-of-War 'Wampanoag,'" in* Cassier's Magazine, *for August and September, 1900. Left, the general plan of the ship's boiler and engine rooms. Upper right, Longitudinal view of the main boilers. Lower right, cross sectional view of the main boilers. Opposite page, a page reproduced from his article.*

showed that a pressure of 1.5 pounds per square inch of pistons was required to work the unloaded engines, and that this pressure was sensibly constant at all speeds of piston.

The loss of speed by the vessel under sail alone, due to the resistance of the dragging revolving screw, was about one-seventh; that is to say, if the vessel under sail alone, without impediment from the screw, could make 14 geographical miles per hour, she would make 12 geographical miles per hour with the screw uncoupled, revolving, and dragging. This proportion was obtained from experiments made by the writer on the U. S. steamship *Massachusetts*, the screw of which could be hoisted entirely out of water and the vessel propelled by the sails alone. In the case of the *Wampanoag*, the axial speed of the screw due to the revolutions of the screw by the reaction of the water upon it when uncoupled and dragging with the vessel under sail alone, was five-eighths of the speed of the vessel, the "drag" of the screw amounting to three-eighths of that speed. This proportion was sensibly constant at all speeds of vessel.

| | |
|---|---|
| Number of cylinders | 2 |
| Diameter of cylinders | 100 in. |
| Stroke of pistons | 4 ft. |
| Diameter of piston rods (2 to each cylinder) | 7½ in. |
| Net area of each piston | 7809.8023 sq. in. |
| Space displacement of each piston per stroke | 216.939 cu. ft. |
| Waste space in clearance and steam passage at one end of each cylinder | 15.0465 cu. ft |
| Per cent. which the waste space in clearance and steam passage at one end of each cylinder is of the space displacement of each piston per stroke | 6.936 |
| Area of steam port at one end of one cylinder | 567 sq. in. |
| Diameter of the crosshead journal | 15 in. |
| Length of the crosshead journal | 22 in. |
| Diameter of necks of connecting rods | 10 in. |
| Length of connecting rods between centres of journals | 10 ft. |
| Number of main journals for both engines | 4 |
| Diameter of main journals | 18 in. |
| Length of main journals | 4 ft. |
| Diameter of main shaft cog-wheel over cogs | 10 ft. 6¼ in. |
| Number of cogs in periphery of cog-wheel of main shaft | 84 |
| Aggregate length of cogs of cog-wheel of main shaft | 9 ft. |
| Diameter over cogs of pinion on screw shaft | 5 ft. 3½ in. |
| Number of cogs in periphery of pinion on screw shaft | 41 |
| Ratio of the gearing | 2.05 |
| Diameter of screw shaft journals | 18 in. |
| Length of screw shaft journals | 4 ft. |
| Diameter of crank-pin journals | 16 in. |
| Length of crank-pin journals | 27 in. |
| Surface condenser, one in common to both engines. | |
| Number of brass tubes in surface condenser | 7,168 |
| Outside diameter of condenser tubes | ⅝ in. |
| Exposed length of condenser tubes | 6 ft. |
| Aggregate area of exposed surface of tubes, calculated for their outside circumference | 7,037 sq. ft. |
| Aggregate cross-section of half the tubes for passage of refrigerating water, calculated for inside diameter of tubes | 1,047 sq. in. |
| Number of water circulating pumps | 2 |
| Diameter of circulating pumps | 28 in. |
| Diameter of piston rod of circulating pump | 3½ in. |
| Net area of each circulating pump piston | 598.37 sq. in. |
| Stroke of piston at each circulating pump | 2.482 ft. |
| Space displacement of piston of each circulating pump per stroke | 10.314 cu. ft. |
| Number of air pumps | 2 |
| Diameter of air pumps | 34 in. |
| Diameter of piston rod of air pump | 4 in. |

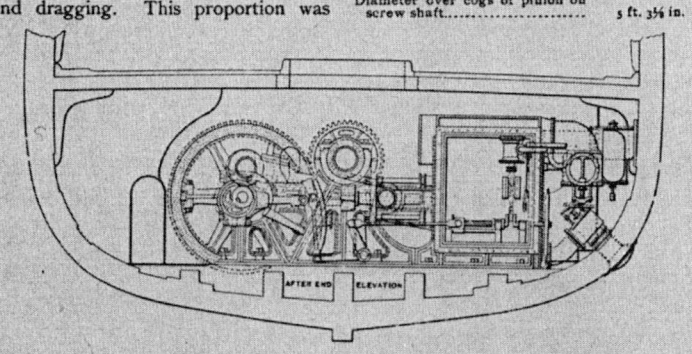

THE ENGINES OF THE "WAMPANOAG"

*Rear Admiral Benjamin Franklin Isherwood, United States Navy.*

construction slowly progressed through 1864 and 1865, as Ericsson and Isherwood exchanged a number of formal letters in which they argued politely about construction details. It was not in Ericsson's nature to be polite, especially not with a man for whom he had no affection, but the inventor was in an awkward position. As Chief of the Bureau of Steam Engineering, Isherwood could easily cut off progress payments being made on the machinery, and Ericsson was too dependent on these funds to risk an open battle.[7]

As a result, Ericsson had to depend on intermediaries to apply persuasion or pressure as it was needed. He chiefly relied on John A. Griswold for such services. Griswold, an iron manufacturer from Troy, New York, had aided in the financing of the original *Monitor;* and from this initial contact with Ericsson, he had become a strong and influential friend. Becoming a member of Congress in 1864, Griswold acted as Ericsson's Washington agent and spent much time in the Navy Department, promoting the inventor's interests wherever possible.

In February, 1864, Griswold called on Lenthall to discuss the Ericsson-Isherwood relationship. Griswold told the naval constructor that he and Isherwood seemed hostile toward Ericsson, that they had not met the inventor's claims with fairness, and that their opposition to him was the source of trouble. Ericsson had not disparaged Isherwood; actually, Griswold maintained, he had done the opposite. When Griswold had, at one time, become "very much prejudiced" against the Engineer in Chief, it had been Ericsson who had tempered Griswold's views by praising Isherwood's abilities.

In answer to these complaints, Lenthall quietly told Griswold that he had never felt any hostility or prejudice toward Ericsson. Unconvinced, Griswold reported to the inventor that it was his position on the House Naval Affairs Committee which had forced Lenthall "to propitiate us [rather than] defy us."[8]

By the end of 1865, Griswold was more than ever convinced that Isherwood's opposition was intractable. Complaining to Welles of Isherwood's and Lenthall's attitude, he argued that they should be more agreeable toward the inventor of the universally accepted monitor system, if only as "some little atonement . . . for their past course" in resisting the ironclad scheme. Writing to Ericsson, October 17, 1865, Griswold predicted ultimate success for their cause, but he had few illusions about its immediacy:

The slow moving finger of justice will I suppose in course of time bring itself over the point of final adjustment—I don't mean Mr. Isherwood's finger—& we shall all breathe an atmosphere untainted by the odor of *disputed claims.*[9]

Disputed claims, however, seemed to be inevitable as the *Madawaska* neared completion. The more Ericsson considered his vessel, the more convinced he became that she was without peer, and the more intolerant he became of bureau criticism. Promising Fox that the *Madawaska* would be the fastest screw warship in the world, certainly faster than Isherwood's *Wampanoag,* Ericsson impatiently waited for the competition between these "hornets," as the English now called them.

Isherwood, however, appeared to be avoiding Ericsson's challenge. He seemed totally unsympathetic to Ericsson's financial situation, forcing the inventor to use every cent of his own funds to supplement the progress payments on the *Madawaska* which Ericsson pried with difficulty out of the steam bureau. These payments, by the end of 1865, had become an obsession with Ericsson, and he insisted that the Engineer in Chief was arbitrarily withholding funds and was being "palpably unjust."[10]

In early 1866, Griswold went to the steam bureau offices in an attempt to reduce the amount of funds withheld from progress payment on the *Madawaska* construction from $100,000 to $50,000, but he found Isherwood unwilling to reduce this amount until Ericsson satisfactorily completed the terms of the contract. Griswold then became convinced that Isherwood was deliberately dragging his feet and was discriminating against Ericsson, since the inventor, unlike the contractors for the *Idaho* and the *Chattanooga,* was not obligated to meet performance requirements, but was simply to satisfy the requirements for good workmanship.

Writing to Ericsson, April 3, 1866, of Isherwood's decision, Griswold stated that he had had a long talk with the bureau Chief and "should have had an angry one had I not thought it policy to preserve a show of patience. I only succeeded in *half* doing this," Griswold admitted to his friend.[11] Ericsson considered Isherwood's attitude to be "monstrous"; but on reflection, he, too, recognized the need for discretion, and in answering Griswold he merely remarked that the Engineer in Chief's comparison of him with the other contractors was "very improper."[12]

Looking to Welles for assistance proved fruitless for the inventor. The Secretary told Ericsson that he could not reduce the amount of money reserved by the government because of Isherwood's reiterated opinion that $100,000 was "not too much to secure the government interest."[13] Ericsson then despondently wrote Griswold that he had lost hope after Welles's refusal to give them assistance. Nevertheless, he encouraged Griswold to visit the obdurate Engineer in Chief once more, in order to point out that Ericsson was in a different category from the other builders, and that in patriotically refusing to delay making his contract at a time when prices for materials were high, Ericsson had handicapped himself in comparison to the other contractors.[14]

The problem, Ericsson explained to Fox, was that Isherwood was "one of the most unfair persons in the engineering profession, who from his position can get any number of engineers to sustain any condemnatory report which his fertile pen may draw." Since Isherwood had already decided that Ericsson's vibrating-lever monitor engine was worthless, the inventor asserted, there would be great difficulty in obtaining any more money from the government, let alone in getting the *Madawaska*'s machinery accepted. Well might he "live in dread," Ericsson complained, when the Engineer in Chief first insisted on a 144-hour trial as an indispensable condition to acceptance of the machinery—and then told Griswold that such a condition was impossible to fulfill.[15]

By May, 1866, Ericsson had exhausted his patience with the Engineer in Chief. "My object in asking you to see Isherwood," he finally informed Griswold, "is not that you should be the 'recipient of his consideration,' but that *he* should be the *recipient* of some good solid words from you on the subject of his gross injustice to the Madawasca contractor. . . ." Insisting that the Engineer in Chief had "committed a damnable piece of injustice in my case," Ericsson urged Griswold, "Do not be afraid to speak out—you cannot possibly produce a worse state of things than now exists, but you may by strong language do much good—. . . ." Astonished that the Navy Department had refused to take charge of the now completed machinery for the *Madawaska,* Ericsson concluded that there was villainy intended by the bureau Chief.[16]

In his correspondence with friends in England, Ericsson gave full vent to his frustrations with the Bureau of Steam Engineering.

Accusing Isherwood of having been his persecutor for twenty years, the inventor branded him as stupid, villainous, utterly devoid of constructive skill, and a good politician rather than an educated engineer. Isherwood might be clever as a writer and compiler, Ericsson allowed, but he had nothing else to demonstrate, except the dubious distinction of being "dishonest and the greatest scoundrel I have ever known."[17]

Much of Ericsson's irritability was caused by the imminence of the *Madawaska*'s trial. For the inventor, the success of his engines had now become vital. In agony, he realized with increasing clarity that the man who controlled the testing, the man who would decide whether or not the *Madawaska* was to be a success, was the very man whose efforts and abilities the *Madawaska* had been designed to ridicule.

By the middle of June, Ericsson believed "the crisis is so near at hand and the danger so imminent" that he urged Griswold to protect his interests against Isherwood's "deep scheme" to effect the *Madawaska*'s failure. The Engineer in Chief, he warned, would order the 144-hour trial without previously trying out the engines or drilling the crew of engineers. In this way, "the vile official" would justify his withholding of the $100,000 by insuring that the vessel would be condemned by his board of inspecting engineers.[18]

Although Isherwood had sent him instructions to use his own engineers in making the engine trials, Ericsson refused to accept this opportunity to protect his interests. Isherwood's superheating boiler arrangement was so peculiar and "fraught with great danger," Ericsson insisted, that no engineer should operate the boilers who did not have extensive experience with Isherwood's machinery. By suggesting that Ericsson use his own men, Isherwood was actually contriving to destroy the vessel, for running the machinery at full speed, without first breaking it in at the dock, and using an untrained crew "would inevitably lead to its destruction," Ericsson maintained. If he used his own crew, Isherwood's boilers would undoubtedly explode, yet with Isherwood's engineers running the ship, the outcome of the trial was foreordained.[19] "I enclose copies of my statement and protest against the intended destruction—('trial')— of the Madawasca," Ericsson dejectedly wrote Griswold.[20]

Isherwood tolerantly listened to the objections flooding in from the inventor. Ericsson, he decided, could have as many preliminary trials of his vessel as he wished before the final 144-hour official test. He could run the ship at the dock or at sea, with any boiler pressure and at any speed, and he could omit the objectionable superheater if he chose. Of course, Isherwood added, Ericsson would have to pay for all of these preliminary trials, since they were not a part of the contract. The "terrible accidents" depicted by Ericsson existed only in his imagination, Isherwood maintained. No marine engineers were so incapable as to be unable to work the simple superheating system, which, in fact, had been in use for over 1 year on 8 warships and had given no trouble. "Nothing will be spared on the part of the Bureau," Isherwood promised, "to enable the Madawasca's machinery to prove the great success the Contractor and the Navy Department desire it to be."[21]

The equanimity with which Isherwood had greeted the fulminations of Ericsson could be easily explained. The Engineer in Chief was not a vindictive man, and it would have been totally uncharacteristic of him to eliminate competition by forcing his opponent's machinery to fail. Isherwood was too convinced of the superiority of his own designs to resort to an underhanded elimination of competing ones, and he had seen too much of Ericsson's work to fear the results of any competition. The test of the machinery would be in practice; and as a practical man, Isherwood recognized the weakness of Ericsson, the inventive genius who so often ignored the little details which might make all the difference in actual operation.

Ericsson and his associates, however, perceived another reason for Isherwood's surprising mildness at a moment when they expected him to be most harsh. Convinced that Isherwood still desired Ericsson's ruin, they soon found a suitable explanation for the Engineer in Chief's "friendliness." Four years had elapsed since Isherwood had been first nominated as bureau Chief; by law, his term of service in that office had expired and he was now up for renomination.

A man who had freely made enemies by simply doing his duty, Isherwood now needed all the help he could get to retain his post. "He is, of course, desirous of conciliating all that might interfere with his confirmation," Griswold wrote Ericsson, adding that the inventor could now have his own way in the trial of the *Madawaska*. On the other hand, Griswold conceded, a successful trial and the recognition of Ericsson's vessel as superior to any existing government vessel would seriously damage Isherwood's prestige and, therefore, his chances for renomination. This possibility, Griswold admitted, might cause the Engineer in Chief to insure the relative success of his own vessel, even before its trial, by forcing the *Madawaska* to fail.

Measuring these possibilities, Ericsson, on July 13, 1866, finally came to a decision and wrote to Griswold:

> I hope you will not hesitate to exert your utmost influence to turn out the dishonest Bureau Chief, as there is not the remotest chance of his ever accepting the Madawasca's machinery—I have almost made up my mind to write a letter to the Secretary of the Navy showing the imperative necessity of removing Isherwood as the only means of saving the great screw navy, now being built, from utter failure. . . .[22]

However, Ericsson's delay in arriving at this determination to oppose Isherwood was fatal to his cause. Andrew Johnson had nominated Isherwood as bureau Chief, on June 28, to fill a four-year term, beginning July 23, when his present commission expired. The nomination was presented to the Senate on July 3 and was immediately referred to the Committee on Naval Affairs. Despite widespread hostility toward Isherwood, there was no organized opposition to his renomination; and on July 12, Senator Grimes, of the naval committee, reported favorably on it to the Senate. On the following day, Isherwood received the Senate's confirmation, and he continued his duties uninterrupted as Chief of the Bureau of Steam Engineering.

On July 15, Griswold wrote to Ericsson, reporting that he had heard of the confirmation, but was not sure that it was true, although he feared that they were too late in their attempt to block the renomination. Informing the inventor that he had spoken to a number of men about opposing the bureau Chief, Griswold doubted if they could have defeated Isherwood in any case.[23] Within a few days, Griswold presented definite information to Ericsson, relating that Senator Grimes had told him that there had been no opposition at all to Isherwood in the Senate, although the Senator would have fought the renomination had he received any support. As a result, Griswold regretfully concluded, nothing could be done about the Engineer in Chief until the next session of Congress.[24]

With Isherwood firmly settled in the office of bureau Chief, Ericsson grimly resumed his efforts to save the *Madawaska* from the impending doom which the inventor was convinced was about to befall his cruiser. The other contractors, he informed Griswold, were in as difficult straits. As Isherwood had set "impossible" conditions for all of them to fulfill, Ericsson concluded that the Engineer in Chief "is determined not to accept any engine not built to his own plans." The *Madawaska* would be accepted, Ericsson assured Griswold, but not until Isherwood had ceased being bureau Chief.[25]

The preliminary trials of his cruiser only substantiated Ericsson's fears. A tremendous antagonism had risen between the naval engineers on board and his own engineer. Ericsson finally had been forced to remove his representative, Isaac Newton, who was so thoroughly despised by the others that Ericsson was afraid that they would deliberately ruin the machinery just to spite him.

Isherwood, however, was not satisfied with this arrangement, realizing that with no representative of Ericsson on board the vessel the inventor would have too easy an excuse to protest any trial results. He therefore wrote to Admiral Gregory in New York, asking his assistance in convincing the stubborn inventor that he should put his man back on board.[26] Ericsson shrewdly replied to Gregory that he would not do so unless Isherwood eased the trial requirements and withdrew his insistence on a 144-hour trial run.[27]

After several weeks during which Ericsson protested about the length of the run, the dangerous boilers, and the motives of the Engineer in Chief,[28] the *Madawaska* finally had her preliminary engine trial. The results were inconclusive. Ericsson's engine failed to work off the steam generated by the eight large boilers, and the journals, where shafts turned in their bearings, overheated, forcing the engine to stop running after thirty-three hours of operation. Ericsson declared that the fault lay in Isherwood's insistence on too much boiler power, but the Engineer in Chief, in two long letters, explained with remarkable restraint and in great detail that the fault lay with the journals. To mollify the inventor, Isherwood offered him another trial in which he could use his own men rather than Navy engineers. Ericsson at once concluded that Isherwood had "caved in," that the *Madawaska* was a great success, and that the Engineer in Chief now was only concerned with preventing a race between his ship and Ericsson's.[29]

Isherwood undeniably was taking pains to satisfy the inventor's demands. On November 11, he urged Ericsson to put as many of his own men as he desired on the *Madawaska* and have them operate the engines "in all respect according to your directions." The steam bureau, its Chief promised, would give Ericsson all the assistance he required.[30] On December 4 and 7, Isherwood again wrote to Ericsson, assuring him that the engines would not be destroyed, as the inventor had assumed, and that the 144-hour sea trial was neither unusual nor onerous for such a steam vessel.[31]

The probable explanation for Isherwood's solicitous attitude was his desire to complete the *Madawaska*'s trials. Instead of avoiding a competition, as Ericsson believed, the Engineer in Chief was simply tired of his tedious negotiations with the intractable inventor. As Isherwood pointed out to Fox, the *Madawaska* had been through a series of "abortive attempts" at performing at a fifteen-knot speed, but so far had been able to make only eleven knots in smooth water over a twenty-four-hour period, and then the machinery was so injured that it would take two months to prepare the vessel for another trial.[32]

Although Isherwood patiently humored Ericsson, he could not remain imperturbable toward the continual attacks coming from Ericsson's loyal booster, William Church. After several years of ignoring the holier-than-thou judgments rendered against him, Isherwood finally decided to reply to a particularly annoying editorial of November 3, 1866, which had pronounced his engines a total failure. First excusing the *Army and Navy Journal* for having been imposed upon by its informants, Isherwood then turned on Church, insisting that he was as guilty as Ericsson, Newton, and others who were the *Journal*'s source of information. "Indeed," he concluded, "the number of falsehoods in your editorial is about equal to the number of its statements, while the opinions expressed parallel inaccuracy and malevolence and invented facts; and, I regret to add, that those remarks apply with too much truth to all the comments which have appeared in your journal on [my] machinery...."[33]

Such righteous indignation fell on unsympathetic ears. On November 24, in a patronizing editorial printed in the same issue with Isherwood's letter, Church first chided Isherwood for his lack of calm judgment, so necessary for his position of responsibility. The *Journal* had no desire to do Isherwood injustice, he continued, and it would certainly keep silent were it not convinced that Isherwood was leading the Navy to ruin by using "power and official patronage" to sustain his fallacious theories. So infamous had these theories become that the most telling judgments on the Engineer in Chief's work were often pronounced by foreign journals. With this in mind, Church wrote an editorial on December 29 in which he turned, as he so often did, to the London *Engineer* for a suitable *coup de grâce* to the Engineer in Chief:

> He may perhaps find some consolation in knowing that in the old country little or no sympathy is felt with those who would wish to see his post taken by another. On the contrary, we believe him to be the right man in the right place. Indeed, we could wish to see his principles and his practice adopted by every naval power in existence—except Britain.[34]

After years of optimism, accusation, and frustration, the final testing of the *Madawaska*'s machinery took place on January 14, 1867. The vessel went out for her 144-hour trial in a rough sea which grew steadily worse as the wind rose in intensity from force 7 to force 11, a true hurricane. By early morning on January 17, the steamer was making 15 knots, but she had to slow down, as the wind changed direction and blew steadily in the face of the vessel. With her steam cutoff set at $3/10$ of the piston stroke, the ship made her 15-knot speed with only 20 pounds of steam pressure from her 8 boilers, 10 pounds below normal operation.[35]

As the seas grew worse, the vessel ran into difficulty. Although she was able to attain fifteen and one-half knots for brief periods, the cruiser could not maintain this speed because she rolled badly and had so little buoyancy that the seas rolled into the ship. Squatting down into the oncoming waves instead of riding over them, the *Madawaska* was in continual trouble, with the engine room flooding several times while her crew grimly believed that the ship was about to founder.[36] Unable to steam for any length of time, the cruiser used all of her coal while staying at sea for one week, battling the heavy seas and steaming for one thousand miles. In order to establish trial data, the board of examining engineers had to combine the three best periods of performance in order to aggregate even a forty-one-hour total, and it was on this data that the *Madawaska*'s performance was judged.[37]

Although making over 15 knots for several short periods of time, the vessel could average only 12.732 knots for the full 41 hours, while burning coal at the inefficient rate of 5.170 pounds per horsepower hour. On this basis, the *Madawaska* had failed to meet the speed requirements for such a commerce destroyer. It was apparent that Ericsson's machinery was not entirely responsible for this substandard performance. The ocean had been unusually rough, and the vessel's captain reported that he had never seen a ship "take seas into her so viciously," exhibiting a "hard-weather helm" which challenged the best of sailors. "Her machinery," the captain insisted, "is as near perfect as it need be. It has undergone the severest test and not once been found wanting."[38] The only mechanical trouble, he maintained, had been at the start of the trip when the coupling bolts had worked loose and the vessel had to be stopped in order to recouple the propeller.[39]

The main difficulty appeared to be in matching the machinery to the hull of the vessel. Ericsson's engines, despite Isherwood's later denunciations,[40] were basically sound in design and capable of performing reliably when placed in a proper vessel. However, the *Madawaska* was of an unusual design. With speed the primary consideration, she was extremely long for her breadth, and with her wooden hull she presented a difficult challenge to any engineer who attempted to fit in the heavy and powerful engines necessary to drive her at great speeds. Necessarily lacking in rigidity, the hull of the cruiser placed enormous strains on the propeller shaft, and the hull was not sturdy enough to take the tremendous pounding that Ericsson's large, direct-acting engines delivered. Thus it was the right engine but in the wrong ship. Ericsson's machinery might drive the *Madawaska* for brief periods of time at high speeds, but for continual operations it was a failure.

The Navy Department accepted Ericsson's engines, despite their inability to achieve a continual fifteen-knot speed, but the acceptance was a hollow victory for the inventor. His pride was hurt when the *Madawaska* averaged less than thirteen knots on her trial. Accusing Isherwood of reporting only the average speed in an attempt to gloss over the top performance of the cruiser, Ericsson believed that the Engineer in Chief was currently very embarrassed while trying to rationalize the unexpected success of the *Madawaska's* machinery. The cruiser, after her "triumphant return," had been immediately taken out of commission, and it was obvious, the inventor declared, that she would be dismantled as quickly as possible in order to avoid a ruinous comparison with Isherwood's ships. Not only did

Isherwood now see that his cruisers could provide only a pitiable challenge to the *Madawaska,* but the Secretary of the Navy, Ericsson announced, was trying to hide the fact that his Engineer in Chief was designing a "whole fleet of ships" which would all prove inferior to that powered by the machinery of an outside engineer.[41]

With the *Madawaska* in the government's hands, all that remained was the final settlement of the contract payments. As might be expected, there was an immediate disagreement over who should pay for the various trials, Isherwood at first maintaining that the steam bureau was liable only for the costs of the final speed trial and that all the preliminary trials which Ericsson had demanded should properly be paid for out of the builder's pocket. After a series of letters between Ericsson, Griswold, and Isherwood, the Navy Department, on the bureau Chief's recommendation, decided that Ericsson did not have to pay for the coal used in the preliminary trials. For once, Griswold admitted to the inventor, Isherwood had not stood in their way.[42]

With the conclusion of the *Madawaska* issue, the formal business relationship between Isherwood and Ericsson ceased. However, this was not the end of the prolonged and bitter controversy over their respective engineering merits. The *Madawaska* had been designed expressly to compete with ships powered by Isherwood's machinery, and although these vessels were still building, they would soon be completed and have their speed trials. Then the issue would finally be resolved. Ericsson, in 1867, was sure that the matter had already been settled. His machinery, he believed, had demonstrated enough power, although in brief and sporadic operation, to withstand the challenge of any engines the Engineer in Chief might contrive to drive the sister ships of the *Madawaska.* Isherwood, equally sure of himself, waited impatiently until his engines could vindicate his theories and designs.

In the meantime, the *Army and Navy Journal* never ceased harassing the Engineer in Chief. Convinced that Isherwood had maneuvered circumstances so that the *Madawaska* would fail to produce sufficient speed, Church heaped scorn and ridicule on the bureau Chief's head. More than ever he was certain that Ericsson's genius was unassailable and that the Engineer in Chief was an utter incompetent. The trial of Isherwood's engines, he promised, would give sufficient proof of that.

# IX. Isherwood's Masterpiece

In the years following the Civil War, Isherwood found his efforts plagued by a new hindrance—economy. Although the war had conclusively determined that naval power would thereafter be dependent on steam, the need for a large Navy for the United States was no longer so apparent. The great cost of the war produced a postwar reactionary desire for retrenchment in naval expenditure; and, by slashing appropriations, Congress soon demonstrated its lack of sympathy for the ambitious and fruitful projects of the Navy's Engineer in Chief.

When Welles proposed to modernize the Navy by enlarging and improving Navy yards, and by developing large iron-manufacturing facilities, he received occasional support in Congress from those who realized that industrial power now held the key to naval strength. Nevertheless, the money needed to realize these projects was not forthcoming. Within a few months after Appomattox, the Navy Department was in the grip of an economizing drive which weighed with particular severity on the Bureau of Steam Engineering, whose operations had so greatly expanded during the war.

By the end of 1865 the work of his bureau had slowed to the point where Isherwood reported to Welles, October 12, 1865, that he required no appropriations for the next fiscal year, since a large surplus remained from the previous year's funds. The stringent restrictions forced upon his bureau's operations had stopped all new construction and had strung out work in progress so that he estimated that the balance of funds should last for two years or more, at the present rate of expenditure.[1]

At the same time, Welles requested a monthly report from Isherwood which would itemize in detail all the steaming henceforth done by naval vessels. This report was to list the amount of coal consumed, the duration in hours each time a vessel raised steam, and the reasons for each use of the engines.[2] Collecting and organizing data for these monthly reports became a regular part of Isherwood's office routine, as the need to save on fuel increasingly determined the operations of the Navy. As Senator Grimes explained to Congress, "Our vessels are instructed to sail all the time except during a storm or going in or out of port." They were not ordinarily to use coal because it was too expensive, and besides, the use of sail power was supposed to produce better sailors and officers.[3] What the Senator did not bother to explain was the "economy" derived from placing on board the steamer a full complement of sailors along with all the coal heavers, firemen, and engineers needed to operate such a vessel.

At the end of the following year, Isherwood again made no request for money for bureau operations, although he recommended considerable salary increases for his small office staff. His expenditures were at an absolute minimum, he informed Welles. He had started no new work, his force of mechanics had been reduced to the smallest practicable level, and a number of engines already building would be taken unfinished from the contractors and stored, since the construction of vessels which were to have used them had been postponed indefinitely.

Finally compelled to resume his requests for money when the bureau's funds were exhausted, Isherwood still continued to economize. At the beginning of 1868 he reported to Welles that, within one month, he planned to reduce the bureau's civilian work force in Navy yards from 1,700 to 884 men. This cutback had become especially necessary, he explained, since Congress had granted him an "excessively small appropriation" and then had passed a law reducing the workday for government employees from 10 to 8 hours while maintaining the same total pay. The shorter workday had forced a reduction of 20 per cent in the amount of work his money could hire, complained Isherwood, in a report of October 22, 1868; and it meant that he could make virtually no repairs on the machinery of steamers coming in from cruises, but only "protect it from further deterioration."[4]

The shortage of funds brought vigorous protests from the Engineer in Chief when it affected the testing of his naval machinery. Sea trials of the few new ships put into commission were delayed, and when they did occur, the lack of money permitted only half the normal complement of firemen and engineers for trial runs. With a shortage of coal, vessels ran using only half of their boilers, and they were unable to achieve their designed speed. Even so, Isherwood noted with momentary satisfaction, one of his large sloops of war, the *Contoocook,* made thirteen knots in smooth water at a reduced consumption of coal. When using only half of her boilers and no superheaters, she could still make ten knots against a strong head wind and a heavy sea.

Although Isherwood often could not find a ship in which to put the engines under construction, he strongly argued against halting work on the machinery before it was completed. Ships could be built in a few months, he explained, but the engines took much longer, and it was wise to have enough of them in storage for one dozen vessels. With this in mind, he carefully hoarded the meager funds allotted to his bureau and worked busily to complete as many engines as possible, particularly before March, 1869, when the incoming administration might not be so tolerant of his policies.

While many of his engines had to be stored, those of Isherwood's design which were in service did great credit to the Engineer in Chief. During the construction of the small class of gunboats of the *Nipsic* class in 1862 and 1863, five of the eight vessels had received machinery designed by other engineers. One of these gunboats, the *Kansas,* had engines of English design and manufacture which had been captured on their way to a Confederate port where they were to have been used to power a rebel ram. After the war was over, an opportunity to compare these engines with Isherwood's arose when the *Kansas* encountered her sistership *Nipsic* while on a cruise during January, 1866. After two days of trial, in which the gunboats raced for twelve consecutive hours while burning the same amount of coal per hour, the *Nipsic,* with Isherwood engines, averaged eleven knots to the *Kansas'* eight.

The English engine ran on the principle of high expansion, with a cutoff at one-fourth stroke, but the boilers proved to be inadequate by providing only twenty pounds of steam pressure. Isherwood's machinery on the *Nipsic* used a six-tenth cutoff, and, with thirty-five pounds of pressure, it vindicated the Engineer in Chief's policy of using small engines with large boilers which could supply all the steam required.

A similar trial between Isherwood's engines and competitive English machinery occurred as a result of Assistant Secretary Fox's having previously contracted for English engines for the *Quinnebaug,* a small screw sloop launched in 1866. This vessel in her trial in New York Bay made seven knots, using a cutoff at one-fourth stroke and employing small boilers for her engines. Her sistership *Swatara,* with Isherwood engines, made twelve knots on her trial, with a cutoff at six-tenths and burning the same kind of coal as had the *Quinnebaug.* On the basis of their comparative speeds, Isherwood calculated that his machinery had actually developed two and three-fourths more power than had the larger English engines, since the amount of additional power required increased at a rate proportional to the cubes of the relative speeds.

The engines which Isherwood had originally intended for the *Quinnebaug,* before Fox decided to send abroad for the machinery, were sent to the Naval Academy in 1866, where they were put on display in the new Steam Engineering Building at the school. Vice Admiral David D. Porter, superintendent of the Naval Academy, declared that Isherwood's "beautiful propeller engine" was not only fine for instructing the midshipmen, but was, in addition, unexcelled in design. He was pleased, the Admiral added, to see this "monument of the skill and perseverance of . . . Isherwood" installed at the Academy where it would stand as a permanent refutation to Isherwood's detractors.[5]

The consistent success of his engine in competition with other models brought Isherwood great satisfaction. This success was, in great part, he realized, the result of his insistence on practicality in engine design. "It has been an immense advantage to me in designing machinery," he explained to Fox, October 14, 1867, "to have had a long practical experience in running it at sea, so that I knew exactly what were the real practical conditions which control success. . . ." With pride, he announced that he had not succumbed to the "foolish notions of shop engineers, who, though they may be tolerable mechaniciens [*sic*], cannot be trusted." In this class of "shop engineers . . . who really have no knowledge of what is required," Isherwood included virtually every well-known civilian engineer, such as Merrick, Bartol, Ericsson, Coryell, and Copeland, most of whom at one time or another had unsuccessfully challenged the theories and designs of the Engineer in Chief.[6]

Although the economy drive in the department seriously curtailed the building and repair of machinery, Isherwood was more successful in maintaining the experimentation which was so close to his heart. The use of petroleum as a replacement for coal particularly intrigued him, since there was an abundance of oil in the United States, it took up less weight and space, it required a much smaller engine crew, and fires could be started and stopped at once instead of spending the hour required to build up a steady coal fire and the hour needed to burn it out.

The initial experiments in 1863 had been disappointing, but in 1866, Isherwood began again a full-scale testing which would determine the feasibility of oil as a steam-engine fuel. By December he suspected that, once again, the results would be discouraging; but he persisted in completing the work. By May, 1867, he reported to Fox that all the experiments in New York with oil had "proved complete failures," since the petroleum, to everyone's astonishment, had evaporated less water than a comparable weight of anthracite coal. In addition, burning the oil left an excessive amount of coke and tar deposits which clung with such tenacity to the iron furnaces that the heating apparatus was ruined and had to be discarded.[7]

All of the engineers connected with these trials had agreed that crude oil appeared to be a failure as a practicable fuel for marine engines, Isherwood informed Fox, since the Engineer in Chief's experiments, as ever, had been exhaustive. The more he had experimented, the worse conditions had become, for the smell was unbearable, there was a danger from volatile gases, and what little economies resulted from any reduction in bulk and weight could not compare with the disadvantages. Discouraged, the Engineer in Chief summarized his findings to Welles:

> The use of petroleum as a fuel for steamers is hopeless; convenience is against it, comfort is against it, health is against it, economy is against it, and safety is against it.[8]

The technology of the 1860's had proved inadequate, and it would be nearly half a century before warships would be converted from coal to oil.

Isherwood had hoped to publish the third volume of his *Experimental Researches* after the war, but he was doomed to disappointment. Although he had the manuscript ready, there was no money forthcoming from the engineers who had previously supported him. Attacks on his published work had increased in proportion to the mounting success of Isherwood's machinery in comparison with those of other engineers. The daily newspapers continually ridiculed the Engineer in Chief for his theories, and with the drive for economy in the government, he became a symbol of irresponsible waste. His "anthracite sacrifices" in making boiler experiments were deplored, while *The New York Times,* on January 11, 1866, dismissed the dock trials of his machinery as mere "baby-house play of burning coal." His several volumes on engineering they regarded as "transcendental trash" which had been forced upon unwilling naval contractors and naval engineers.[9]

So often alone in a hostile engineering world, Isherwood welcomed the encouragement of Gustavus Fox. While serving as Assistant Secretary, Fox had often opposed the theories of the Engineer in Chief, especially over ironclads. Nevertheless, he had recognized the value of Isherwood's scientific work and had given him the support he needed. But for Fox, Isherwood maintained, the hundreds of experiments on boilers and on the use of expanded steam would not have been made. When finished, these experiments would stand, Isherwood declared, as "the most valuable mass of information ever given to the world in engineering." Technology had advanced more during the Civil War than during the previous half century, the Engineer in Chief noted; and it was Fox, "the man at the head of affairs who believed in progress," who had demonstrated the energy and had authorized the money which made possible such progress. "I often wonder," Isherwood reflected, "when I consider how general and violent were the attacks on me, that the Dept. should have sustained me as it did. . . ." Fox, he knew, was the main reason for this support, and now Isherwood could repay him for his confidence by producing experimental results which would live long after "you who ordered them & I who executed them, are forgotten."[10]

At the moment of his greatest achievements in engineering, a tragic note sounded in the life of Benjamin Isherwood. John Lenthall was seriously ill, and no one held any hope for his recovery.

Lenthall had been an intimate friend of Isherwood for years, and their close association throughout the war had furthered this friendship. Indifferent to his wife and children, Isherwood had made naval engineering his life; and Lenthall, so close to him in his work, held the unbound affection and respect of his younger friend.

Many others recognized the qualities which Isherwood venerated in the sedate, elderly naval constructor. Senator Grimes had proclaimed to the Senate that Lenthall was "one of the most competent and one of the purest men . . . that I have ever had the fortune to be acquainted with," and he related that one eminent naval officer had declared that " 'if he had lived under any other Government he would long since have been knighted for the services he had rendered to his country.' "[11]

Isherwood respected Lenthall as much for his professional abilities as for his character. Writing to Fox in December, 1866, he agreed with the Assistant Secretary that "when he leaves the world he will take out of it more knowledge of naval architecture than will remain in it—at least in America—He has never been properly appreciated, except by the very few who know him intimately and professionally."[12]

Suddenly, this sixty-year-old gentleman had fallen ill, and Isherwood believed that "sentence of death, to be executed only too soon I fear, has been passed upon him." Apparently, Lenthall had become a victim of "the fatty degeneracy of the heart." Precise even in his misery, Isherwood reported to Fox that Lenthall's heart muscle had been so weakened that it no longer had "sufficient projectile power to send the blood through the system." The naval constructor could barely walk, and then only "at a snails pace," the slightest excitement would "prove instantly fatal," and he was overcome by complete nervous prostration. "It fills me with grief to see him," Isherwood remarked, marveling at the old man's ability to meet his fate "like a stoic," by continuing to perform his normal duties until incapacitated by weakness.

On his doctor's advice, Lenthall and his family were to leave in May for Havana for a rest cure. He would return "in about six weeks if alive, and sooner if dead," Isherwood reported in a letter to Fox dated May 1, 1867, adding solemnly, "I expect to take my last leave of him today, for I have only a bare hope of ever seeing him more." Now that Lenthall was about to depart, presumably forever, everyone suddenly began to recognize the naval constructor's true worth, Isherwood wryly noted, remarking to Fox how Lenthall had become surrounded by those who rushed to aid him in his final days. "He will leave no better and no abler man behind him," Isherwood gloomily concluded, "and I never expect to look upon his like again."[13]

Fortunately, the Engineer in Chief had been too pessimistic.

Whether caused by the salubrious climate of the Caribbean, or by his escape from the stifling heat of the enervating Washington summer, Lenthall began to recover his strength. By late July he had returned to the Jersey shore and was able to move about his house. Greatly cheered, Isherwood suggested to Fox that the naval constructor should take a year's leave of absence in order to travel leisurely through Europe, since he was as mentally acute as ever and lacked only physical strength. Such a voyage, Isherwood reminded Fox, would also be a "professional pilgrimage" which would benefit the government as much as it would Lenthall.

Reflecting on the recovery of his friend, Isherwood was sufficiently moved to exclaim, in a letter to Fox, July 26, 1867, "So much mental power, so much professional experience, so much honesty—in short, so much of all that is best in morals and intellect are rarely found in one person."[14] It was these values, of paramount importance in Isherwood's own life, which had won for John Lenthall a loyalty and devotion few would ever share.

While the economizing drive gripped the Navy Department ever more firmly in the postwar years, Isherwood stubbornly fought to preserve the strength of the steam Navy. Encouraged to use sail at every opportunity, line officers in command of cruising warships became exasperated with the large amount of space and weight which machinery and coal occupied in their vessels. For a cruise which might last for many months, they insisted on a fair degree of comfort, which meant generous accommodations for officers and plenty of room for provisions. Since they rarely used the machinery, it became a petty annoyance, and they began to question its worth, especially since the large funnels greatly interfered with the sailing qualities of their ships.

Isherwood soon ran afoul of these officers. Placed on a board to examine alterations recently made to the USS *Richmond,* he found that his ideas varied greatly from those of the "sea officers" composing a majority of the board. The new Isherwood engines which the vessel had received increased her speed from six to nine knots, but her fuel supply was still grossly inadequate, the Engineer in Chief maintained. She could steam for only seven and one-fourth days, while in the British Navy vessels of comparable speed could steam for ten days. Moreover, Isherwood insisted, "as we possess neither colonies nor forcing strongholds, it is plainly necessary that our steamers of war should be able to keep the sea longer than those of a power whose drum beat follows the sun around the world."

Instead of concentrating on more room for officers and crew and for provisions, the Navy Department should stress high speed and endurance, he believed. The "power of locomotion and command of time" were the prime considerations of warship design, Isherwood declared, and "it follows that only a portion of the vessel can be appropriated to armament and crew, as remains after the machinery and fuel are accommodated. This condition is inexorable." There could be no compromise in peacetime, he added, because a warship, whether in peace or war, must always be ready for battle. Therefore, her equipment and her performance capabilities must always be up to wartime standards.

The experiences of the Civil War and the current precepts and practices of great European naval powers proved beyond doubt that high speed and great endurance were the "controlling necessity"; and no officer, declared Isherwood, should be so foolhardy as to disregard these dictates. Instead of turning back the clock to the days of slow speed and limited steaming, the American Navy should develop eighteen-knot cruisers of great endurance to meet the challenge of European navies and modern merchant fleets. To do less, he warned Welles, "would prove a national calamity."[15]

While Isherwood presented his impassioned argument for a modern Navy, the "sea officers" threw their weight in the opposite direction. Writing to Welles in April, 1867, the Admiral of the Navy, David Farragut, objected to the "contest now going on between speed, and, I may say, all the other qualities of a man of war." Increasing the size of warships, he complained, only resulted in benefits to the engineers. Crew's quarters and spaces for provisions had been sacrificed for speed, and this was senseless when warships were required to utilize full sail and operate as sailing ships far more than as steamers. Welles, he insisted, must "prevent further encroachments" by Isherwood and his Corps.[16]

Impressed by the arguments of this great naval hero, Welles disregarded Isherwood's pleas for speedy warships and bluntly informed the Engineer in Chief:

Hereafter, encroachments will not be made upon the space ordinarily allotted to officers and men and the other appointments of a man of war, unless it shall appear after careful examination that ample accommodations will remain...."[17]

Writing to Fox, April 25, 1867, the Secretary remarked peevishly that the department was being pestered by engineers who "seem to suppose that naval vessels are built for the purpose of carrying engines and engineers." They had ignored the comfort of the crew, Welles complained, in their "zeal for high speed and powerful motive power." He had often cautioned Lenthall on this, Welles related, but the Chief of Construction had "yielded to some extent to Isherwood," forcing the department to interfere and modify the plans of vessels. He cited as an example the *Pensacola,* whose sailing qualities were nearly worthless because the engineers had insisted on locating the funnel so close to the mainmast that it seriously hindered the use of the sails.[18]

The admonitions of the Secretary and the aggrieved protests of the "sea officers" left Isherwood unmoved. Speed and endurance, he maintained, could not be compromised in vessel design if the American Navy wished to retain its power and fulfill its purpose. Instead of acquiescing to the desires of the line officers, Isherwood determined to display to the nation and the world that no longer could engineering be disregarded or depreciated. To achieve this goal, he relied on the vessel which was to be the culmination of his efforts and his hopes, the *Wampanoag.*

In 1867 the class of Navy Department supercruisers planned four years previously were still in the process of construction, with the *Wampanoag* furthest advanced toward completion. Only the vessels built to compete with Isherwood's machinery had been launched and tested. The first of these, Forbes's and Dickerson's *Idaho,* had been a miserable failure. The second, however, had done considerably better.

This vessel was the *Chattanooga,* whose hull had been built by the reputable firm of Cramp and Sons, in Philadelphia, and whose machinery had been designed and constructed by Merrick and Sons, of the same city. Displacing less weight than the department vessels, the *Chattanooga,* like the others, had been designed to achieve a fifteen-knot speed.

On August 17, 1866, she had her speed trial; and over a twenty-four-hour period, she averaged less than thirteen and one-half knots in a perfectly smooth sea with little wind. Her machinery developed almost three times the horsepower of Dickerson's engines, and she burned coal at a slightly more efficient rate. No serious mechanical malfunctions occurred on the trial run, but the board of inspecting engineers noted that there were many minor problems and an excessive amount of abrasion between metal parts. Isherwood told Fox, December 22, 1866, that the engines were so injured by the twenty-four-hour run that it took weeks to repair them, and it would require extensive alterations and repairs to prepare the vessel for service. After some discussion between the contractors and the Navy Department, the *Chattanooga* was accepted, despite her failure to come within one knot of the required contract speed.

Several months after the *Chattanooga*'s trial, Ericsson's *Madawaska*, identical to the *Wampanoag* except for her engines, had failed to average thirteen knots. Thus, the three vessels designed to compete with those of the department had not been able to achieve and maintain a speed which was necessary for an ocean-going commerce destroyer, although in each case their hull design had been tailor made for extreme speed. The hopes of the Navy now rested entirely on the abilities of its Engineer in Chief.

In the midst of the Civil War when the supercruisers had been proposed, there had been no doubt about their intended function. These vessels were to be commerce destroyers, aimed at the economic heart of European powers that threatened the Union cause by intervention on the side of the South. In 1863, the disparity in naval strength between America and Britain, in particular, was so vast that a full-scale naval competition could not be considered. The only recourse was that which America had taken in the past—attempted neutralization of the enemy's naval predominance by creating havoc among its merchant shipping.

The *Wampanoag* class of cruisers had to be extremely fast in order to prey successfully on English commerce. In order to catch speedy mail steamers, intercept dispatches, break up ocean communications, and disrupt merchant shipping, these warships had to achieve at least a fifteen-knot speed. By so doing they could catch any vessel they wished to fight and run away from any other. To be properly effective in "attacking the enemy's purse," they were to fight only when necessary, keep to sea indefinitely, cruise in the lanes of commerce, and destroy every prize they captured, rather than run the risk of taking it in to the few ports open to them.

Such vessels would have to have enormous engines in order to develop their necessary speed; and yet have full sail power so that they could cruise for weeks, if necessary, without expending fuel except during brief periods of high-speed steaming. As the United States had no coaling facilities abroad, and would be at a particular disadvantage in this respect when at war with Britain, it was, therefore, nearly as important for such vessels to have good sailing qualities as to be able to steam at high speeds.

Restricted by the existing low level of technology and by limited manufacturing capabilities, the designers of these vessels had to use wood, rather than iron in building their hulls. Thus particularly vulnerable to enemy fire, these vessels had to mount heavy armament so that they could fire at long range and avoid close bombardment, in which situation their great speed would be neutralized and their unarmored hulls most exposed.

With this policy as a guide for construction, both the Navy Department and the competing private builders had started on their ships during 1863. Of the vessels with Isherwood engines, the *Ammonoosuc* and *Pompanoosuc* were placed under construction at Boston, the *Neshimany* at Philadelphia, and the *Wampanoag* at New York, alongside of its twin, the *Madawaska,* in which Ericsson was to place his engines.

The *Wampanoag*'s measurements were sufficient to remove all doubts that this vessel was built for speed. Her 335 feet of length provided nearly an 8-to-1 ratio to her 44½-foot breadth. Drawing an average 18½ feet, this vessel was designed to displace 4,216 tons when loaded. Building a vessel this size out of wood, even if it were live oak, involved serious difficulties in design. To prevent the hull of such a large ship from sagging or hogging, both from the weight of its machinery and its own length, the builder had to reinforce it with longitudinal and diagonal stringers of iron plate.

In an age before experimental model basins, hull design was still very much an art, and the Navy Department wisely entrusted the *Wampanoag*'s design to the famous ship designer B. F. Delano. The skill of this shipwright was all important, since the process of design was largely a matter of expert guesswork, the accuracy of which would become apparent only after the ship had completed her speed trials. Delano, however, did not have a free hand in this work. Unlike previous vessels in which the machinery had to be designed to fit into a specific hull, the *Wampanoag* was to have a hull shaped for the machinery already planned by Isherwood. Consequently, the Engineer in Chief took an active part in designing this vessel.

Isherwood originally expected that the *Wampanoag* would be built of iron, because of the unusually heavy weights and strains imposed by his machinery. When lack of construction facilities made this impossible, he necessarily agreed to a wooden hull, but he insisted on a design which made a radical departure from accepted form. Instead of a barrel-shaped midship section for the cruiser, he substituted a square section, approaching the principle of twentieth-century warship design. Working closely with Delano, Isherwood eliminated the old overhanging bow of orthodox warships and substituted a straight stem. The forward water lines of the cruiser showed a decided hollow, giving a general effect, according to a recent writer, quite similar to a modern destroyer.

Delano objected to these radical innovations, by which the ship had been virtually designed around her engine and boiler space. He attempted to modify the lines of the *Wampanoag* into more orthodox dimensions, but Isherwood prevailed upon their friendship to force him to return to the original plans. "It was rather a tight fit," Delano admitted, in a letter to Fox dated September 12, 1863, noting that he had to widen the body eight inches and lengthen the ship several feet to get in all the machinery. Nevertheless, he finally concluded that there was nothing objectionable in this unusual practice of shaping the hull to fit the machinery, since it was obvious to him that the *Wampanoag* would develop great speed from her enormous engines.[19]

Isherwood solved the problem of placing heavy engines in a wooden hull by gearing them to the propeller shaft. In this way, the reciprocations and consequent vibration of the engines would be less, while the propeller could still revolve at high speeds. In a vessel like the *Wampanoag*, in which the hull was so long and narrow that it did not possess much longitudinal strength, there was a problem with direct-acting engines, whose vibrations would loosen fastenings and possibly pound the hull to pieces. Isherwood planned, instead, to gear his engines to the shaft by the use of huge wooden (lignum vitae) teeth, so that one double stroke of the piston would produce 2.04 revolutions of the propeller.[20] Although wood would seem to be a questionable material with which to build engine gears, it was considered practicable in those days, and builders maintained that it helped to deaden the noise of engine operation.[21]

Eight Martin boilers, with a maximum working pressure of forty pounds, were used to drive the *Wampanoag*'s two pairs of simple-expansion engines. To produce greater steam pressure and to retard condensation, Isherwood also placed four separately-fired super-heating boilers of his own design in the vessel. The boilers were well protected from enemy fire by being set sufficiently low in the hull so that the highest part of the machinery was more than two feet below the water line. In addition, coal bunkers lined both sides of the hull to provide an additional buffer. To provide sufficient draft for the furnaces, four large funnels protruded from the vessel. Although Isherwood insisted that they did not interfere with the sailing qualities of his ship, he was well aware of the horror his four "chimneys" caused in a day when one was considered a great nuisance by sailors.

The *Wampanoag* had to have the ability to cruise for long periods under sail alone, and to meet this demand, Delano and Isherwood placed three ship-rigged masts on the vessel. Isherwood objected to the bowsprit, since it obstructed direct fire ahead; and he maintained that there was no need of bowsprit sails, because of the ease of steering a ship of such hull proportions. In addition, the strong, live-oak frame, stiffened by the iron stringers, would make the *Wampanoag* "the most formidable projectile that could be devised when hurled as a ram at her enormous speed against the side of a slow antagonist," according to Isherwood. With this purpose in mind, he insisted on the elimination of this projecting spar. Against the advice of most naval constructors, the *Wampanoag*'s bowsprit was eventually removed after the vessel had been in service.

As an unarmored commerce destroyer, the *Wampanoag* had to have heavy armament, both to stop fleeing victims and to hold off attacking warships. Although there were continual modifications and rearrangements of her weapons, as in most nineteenth-century warships, her normal armament was ten 8-inch, smooth-bore guns and two 100 pounders firing in broadside; two 24-pounder howitzers; two 12-pounder howitzers; and one 60-pounder, rifled pivot gun. The vessel was to carry 106,700 pounds of ammunition, with 150 rounds for the pivot gun and 100 for each of the others. The weakness in this distribution of armament, according to some authorities, was that there was an insufficient amount of fore-and-aft fire. For a vessel designed as a commerce destroyer, several heavy bow chasers would seem a necessity, but the extremely fine lines of the *Wampanoag* prohibited the mounting of a battery of heavy bow guns, and direct fire ahead was restricted to the single pivot gun.

By the end of 1863, construction was well under way on the hull of the *Wampanoag* as she lay alongside of the *Madawaska* in the Brooklyn Navy Yard. The $700,000 engines were taking shape at the nearby Novelty Iron Works, while the boilers had been let out to several contractors in New York city and upstate New York. Initially, the *Wampanoag,* perhaps because she was built in the ship house while the *Madawaska* was building "in the open air," led Ericsson's ship by six weeks, but this advantage of time was not to last. Although the main reason for delaying construction of the *Wampanoag* was the time needed to manufacture her geared engines, there were also delays in the building of the hull. Undoubtedly, a major factor in the latter was the supervision of the Engineer in Chief, who anxiously hovered over his vessel, making certain that no shoddy material or workmanship would go into the *Wampanoag*.[22]

As the cruiser neared her launching, Isherwood's enthusiasm mounted. Like Ericsson, he was convinced that the lines of the hull, so carefully sculptured by Delano, would assure great speed. Unlike Ericsson, however, Isherwood had other vessels of this class to supervise, and his interest in them was nearly as great as for the *Wampanoag*. Writing to Fox, September 12, 1864, the Engineer in Chief declared that the *Ammonoosuc,* being built in Boston, was the finest steamship he had ever seen and could not fail to be a great success. With the same engines as the *Wampanoag,* and with nearly an identical hull, the *Ammonoosuc* had "the power for the speed and the model for the power."[23]

The first of the department vessels to be started, the *Wampanoag* was also the first to be launched. On Thursday morning, December 15, 1864, hundreds of privileged spectators clustered in the Brooklyn Navy Yard to watch the daughter of Captain Case, the executive officer of the yard, christen the vessel. In attendance were Admirals Paulding, Gregory, and, above all, Farragut, the latter having arrived with great fanfare on the *Hartford* the previous day. While "prying visitors" were rigorously excluded from the Navy Yard by the vigilant execution of police regulations, those fortunate enough to observe the launching watched the *Wampanoag* slide into the East River at 11:15 A.M. With an accompaniment of cheering and enthusiastic waving of handkerchiefs, the cruiser "entered her destined element in the most graceful and brilliant style."[24]

Although the *Wampanoag* was now ready for her machinery, construction on the vessel abruptly halted. Isherwood's engines were large and expensive, and by necessarily contracting them out to private builders, the Navy Department suffered the consequences. As the months went by, the engines moved towards completion with maddening slowness, and the end of the Civil War only added to the difficulties. Isherwood had demanded expensive materials, such as brass, and he had presented the builders with extremely detailed plans for his complex engine, which they resented, being used to considerable leeway in their work.

However, the delay in construction was not caused by the Engineer in Chief alone. The officials of the Novelty Iron Works apparently decided that they were going to lose money in building the geared engine, since the contract price would not only fail to provide them with an acceptable margin of profit, but might not even meet construction costs. For this reason, they were reputed to be "unwilling to push work in the completion of which they are sure to be mulcted in quite a large sum," said the *Army and Navy Journal* of March 24, 1866. While the private builders competing with the Navy Department had been able to complete their machinery, work on the *Wampanoag*'s engines thus dragged on interminably.

During the months of delay, and especially as the *Madawaska* came nearer to her trial in early 1867, the *Army and Navy Journal* launched an assault on Isherwood and the *Wampanoag* which increased steadily in intensity. On three successive weeks, just before Ericsson's vessel went out to sea, Church filled his first page with a stinging attack on the Engineer in Chief for committing "an enormous blunder" in building a class of cruisers with geared engines. So needlessly complex was Isherwood's machinery that the *Wampanoag*, Church asserted, was virtually stuffed with iron, leaving no room even for the coal in the hull. The four funnels, he complained, were not telescopic. By sticking "bolt upright nearly as high as the maintop," they required a "forest of wires" to keep them from rolling overboard, and this condition, in turn, made the use of sails almost impossible.

Isherwood's "vilely-planned machinery" was mere rubbish and "fit only for the scrap heap," Church maintained. The Engineer in Chief had not placed bedplates in the hull to anchor his engines and boilers, and the crankshaft was excessively long. The twisting and bending of the long, narrow hull in a rough sea would stop these loosely placed engines by mashing their wooden gears, by binding the journals of the propeller shaft, and by creating enormous friction and heat. The Martin boilers, he continued, were inefficient, the separate superheaters were unnecessary, the surface condensers were inadequate, the fixed, four-bladed propeller would be a tremendous drag when sailing, and the whole arrangement was far too expensive.

The shape of the *Wampanoag*'s hull, Church decided, should allow the vessel to go fifteen knots using machinery of only half the weight, space, and cost of Isherwood's. If the Engineer in Chief insisted upon using machinery of the proposed weight, then the cruiser should go at least sixteen knots, the editor insisted.

Church concluded, in the *Army and Navy Journal* of January 5, 1867, by assuring his readers that this criticism had been conducted purely in "the spirit of scientific inquiry, and with no reference to the moral questions therein involved." He would ignore any "plausible or possible incentive" Isherwood may have had for committing these "costly blunders," Church declared, leaving little doubt that he would dwell on this aspect at another time. His readers knew he would not disappoint them.[25]

By assuming that the hull of the *Wampanoag* was so well designed that the achievement of a fifteen-knot speed would not indicate any special quality in Isherwood's machinery, Church had placed himself in an awkward position. Since the *Madawaska* had the same hull, it stood to reason that the machinery designed by John Ericsson should achieve the same speed without any difficulty. When the *Madawaska* failed by a significant degree to do this, Church had no other recourse than to insist that Isherwood was personally responsible for that cruiser's failure. Consequently, he accused the Engineer in Chief of being the "tyrannical power" who had made sure that the speed trials of the privately built ships would be failures, not to keep from paying the contractors, but because he had to protect his reputation. Therefore, Church reasoned, there had appeared the "star-chamber severity" which ruthlessly destroyed the hopes of private builders in order to preserve the position of the Engineer in Chief.[26]

In the summer of 1867, the *Wampanoag* finally received her machinery, and after minor mechanical mishaps (which Church publicized as "radical defects" in the engines, requiring a material change in their design), the cruiser was pronounced ready for trial. On September 17, she was formally commissioned, and Captain J. W. A. Nicholson took command, bringing with him some of the officers and crew of the *Massachusetts,* which had been decommissioned on the same day.[27]

As the *Wampanoag* approached her sea trial, the criticism of her machinery grew more shrill. After suggesting that Isherwood should subject his own engines to the terrible beating he had given to those of his competitors, Church decided that all the machinery which Isherwood had crammed into Delano's hull should produce a speed of eighteen knots or more. However, he cautioned, no matter what results would appear in the trial reports, they could never be trusted, since no naval engineer would be expected to brave the consequences of condemning the work of his Chief. "If the engines don't absolutely tumble into pieces, we shall of course have a favorable report and a vindication of the genius of Mr. Isherwood," Church remarked cynically, while establishing an excuse in case the *Wampanoag* conceivably did develop a fifteen-knot speed.[28]

Widening his scope to include all of Isherwood's engineering designs, Church insisted that the "palpable professional incapacity" of the Engineer in Chief was such a "subject of grave national importance" that it gave him *carte blanche* to abuse Isherwood.[29] At this point, Church was joined by more reputable authorities on steam machinery. The *Scientific American,* on October 26, 1867, ridiculed Isherwood's geared engines, asserted that he was using them only because his earlier direct-acting engines "had given such wretched results," and strongly condemned Isherwood's "want of engineering skill and common sense" in producing machinery which was "a blunder without parallel of its kind."[30]

At four o'clock in the afternoon of Sunday, December 29, 1867, the *Wampanoag* went to sea on a preliminary trial and returned the following day after twenty-five hours under way. On the following Thursday she made another twenty-four-hour trial run; and on the basis of this steaming, Church reported that the vessel had broken down because of the binding and heating of the journals, and that she would take six weeks to repair. He noted that the vessel had failed to make fifteen knots at any time on the shakedown cruise, and that at one point the engines had reputedly stopped dead with forty pounds of pressure in the boilers. Even though Isherwood had used the best steaming coal available, his vessel, Church declared with pleasure, so far had failed to show to the world the vaunted advantages of the geared engines.

Before the *Wampanoag* departed on her official trial trip, Church determined to fire his parting salvo, and on February 1, he devoted a lead editorial to the "foolhardy boldness" and "folly" of the Engineer in Chief. Isherwood, he declared, had initiated a system of steam engineering which caused most of his profession to stand aghast at his endeavors. His great work, the *Wampanoag,* was obviously a failure, rumored to be laid up with both cylinders split by her "short trial trip in fair weather." Reviewing the stand of his paper over the previous years towards the Engineer in Chief, Church stated:

We speak, and have spoken, from Vol. 1, No. 1 of the *Journal,* to this day, very severely of Mr. Isherwood's practice. But we have done so from mature, profound and complete conviction that he is ruining the Navy by his untenable steam delusions. If we have utterly condemned his theories, we have done so only upon a basis of irrefragable facts.

Pronouncing a conceivably premature sentence of death on Isherwood's cruiser, about to make her decisive speed trial, Church concluded, "We are free to say that there is no such monument of mechanical incapacity as the steam machinery of the *Wampanoag* class to be found in the annals of marine engineering."[31]

At 5:00 A.M., on February 7, 1868, after 4½ years under construction, at a cost of $1,575,644, the *Wampanoag* slowly steamed out of New York Harbor to begin her formal trials while Isherwood waited impatiently in his bureau office for the results. Temporarily disconnecting her propeller from the shaft, Captain Nicholson first tested the cruiser's sailing qualities by running the vessel under topgallants, jib, and course, and with no sail on the mizzenmast. In a smooth sea and with the wind fresh on her beam, the *Wampanoag* made 10½ to 11 knots with no difficulty. She steered well and sailed fast under a fresh breeze, the Captain noted, although she needed to make at least 4 to 5 knots so that the propeller would not act as a drag.[32]

Two days later, the Captain began his steam trials, deciding first to run the vessel at a speed of 11 knots to determine economy of fuel consumption. Although forced to end the trial after 25 hours, because of a heavy gale and topping seas, he had run the vessel for 282.5 nautical miles, using slightly less than 47 tons of coal. With her fuel capacity of 750 tons, this meant that the vessel could steam for 17 days at this speed.

After returning to Sandy Hook to repair minor damages caused by the heavy winds, Captain Nicholson once more took his vessel out to sea on the evening of February 11. This time he determined to achieve maximum speed. For thirty-eight hours the *Wampanoag* knifed through increasingly rough water, steaming southwest towards Hampton Roads, Virginia, while a fresh breeze shifted between her starboard and port quarter, often blowing from directly astern. Despite the heavy seas which forced the vessel to roll more deeply as the hours went by, the *Wampanoag* did not reduce speed. The great engines beat on without pause, and soon every man on board knew that Benjamin Isherwood had realized his greatest hopes. The *Wampanoag* was the fastest ship in the world.

Within the first 6 hours the cruiser's speed had climbed to 17 knots, and then for hour after hour the vessel steadily logged between 16 and 17 knots. For the 38-hour period, the vessel averaged 16.6 knots; and for one hour, made the unheard of speed of 17.75 knots, equivalent to 20.47 miles per hour.

Down in the engine room Isherwood's loyal friend Theodore Zeller had been supervising the exertions of the 23 engineers and the large crew of 180 firemen and coal heavers who labored to keep the giant fires roaring steadily under the *Wampanoag*'s 12 boilers. With the throttle set wide open, the geared engines worked smoothly and satisfactorily, the only difficulty being with the crank pin of the after engine, which began warming during the thirty-eighth hour because a rubber washer had worked loose. Although there was no injury to the engine, Zeller and Nicholson decided to stop the trial at that point in order to replace the defective washer.

Zeller and his crew of engineers recorded that the boilers were developing 31.97 pounds of pressure, which was sufficient to run the engines at 31.06 revolutions per minute and turn the screw propeller at 63.37 revolutions. For 24 consecutive hours, their vessel had made 16.97 knots, and the maximum speed of 17.75 knots had been obtained over 4 separate half-hour periods. With a trained crew of firemen and coal heavers, they maintained, the *Wampanoag* would be able to steam easily across the Atlantic at this top speed.

Bad weather forced the discontinuation of the speed trial, as Zeller decided that the machinery had proved its ability to develop seventeen knots consistently. The *Wampanoag* proudly steamed into Hampton Roads, on the afternoon of February 17, to begin her duties as flagship of the North Atlantic Squadron.

The achievements of the *Wampanoag* included more than her extraordinary speed. By using geared engines, the cruiser had been able to run at a very economical rate. Developing a maximum 4,049 indicated horsepower, the vessel burned 12,671 pounds of mixed semibituminous and anthracite coal per hour, producing a rate of consumption of only 3.129 pounds of coal per horsepower hour at top speed. This amount compared to 5.170 for the *Madawaska*, 6.160 for the *Chattanooga*, and 7.600 for the ill-fated *Idaho*. When steaming on only half boiler power, the *Wampanoag* still made 11 to 12 knots, although burning only 1⅞ tons of coal per hour, compared to 5¾ tons at top speed. At one-quarter boiler power, and burning 1⅜ tons of coal per hour, the vessel could still steam at 9 knots.[33]

At top speed, the *Wampanoag* was able to steam for 5½ consecutive days, covering 2,200 miles. While cruising at 11½ knots, she could cover 4,700 miles. This was more than sufficient steaming endurance for a warship which was to operate primarily under sail, relying on steam only when chasing or escaping from vessels.

Even the cruising speed of the *Wampanoag* was unusual for steamers of the 1860's. The top speed of American naval vessels built before the war had rarely exceeded 11 knots, while they cruised at no more than 9 knots. The fastest British transatlantic merchant steamers averaged less than 12 knots for the crossing, and the rate for ordinary vessels was closer to 10 knots. Without the influence of weather, a medium-sized, well-designed screw steamer could usually attain, at best, a 10½-knot average for an Atlantic crossing, and this achievement was without economizing on fuel, while using sail power whenever practicable. Occasionally, world-famous steamers could do better. The Collins Lines' *Adriatic* made a top speed of 15.91 knots, but over a measured mile using the best Welsh coal, a performance which could not be compared to any extended steaming under normal conditions.

The *Wampanoag*'s speed so far exceeded that of contemporary steamers that it was not until September, 1889, when another American naval vessel, the cruiser *Charleston,* would equal her speed, and no ocean steamer in the world was able to match the 17.75 knots until the *Arizona* achieved it in 1879, over 11 years later.[34]

When he received Zeller's telegram from Norfolk, informing him of the *Wampanoag*'s achievements, Isherwood was jubilant. Immediately writing to Fox, the Engineer in Chief stressed the fact that the vessel had been fully loaded with coal, provisions, and armament; and yet, in a very rough sea, she had attained "the most wonderful success."[35] No other steamer in the world, fully loaded and at sea, could average twelve and one-half knots for forty-eight consecutive hours, he declared. In addition, the performance of his vessel in relation to those of her competitors should be judged on the ratio of the cubes of their speeds, so that in comparison with the "miraculous" speed of the *Wampanoag,* boasted Isherwood, "the *Chattanoogas, Madawascas & Idahos* are simply insignificant." For the Engineer in Chief, the trial of the *Wampanoag* had been nothing less than "the most wonderful performance ever made in Steam Engineering."

Much of the credit for the *Wampanoag*'s success, Isherwood insisted, should go to Fox. "I am both glad & proud," he stated in a letter to Fox, February 18, 1868, "to be thus able to vindicate myself & your confidence in me, which notwithstanding the unprecedented abuse and calumny that interest & envy has heaped on me & are still heaping, never faltered, but with wonderful magnanimity and judgement sustained me throughout."[36] Sincerely as he intended his praise, it must not have escaped the Engineer in Chief that such generous sharing of the credit might provide him with a powerful ally in his disagreements with the "sea officers" who had already raised such objection to his work.

The more he reflected on the performance of the *Wampanoag,* the more wonderful it appeared to the Engineer in Chief. Aside from the dazzling speed, the engines had shown magnificent endurance at top speed, he noted; and the economy of operation was admirable. There was now no doubt in Isherwood's mind about the potential of his vessel as a commerce destroyer:

> Only reflect on the damage this single ship can inflict on English commerce. She can sail over to the British Channel & lay off Liverpool or London with absolute safety, destroying every mail and merchant steamer & other vessel in her vicinity. It would be useless to send against her 100 three deckers going 8 knots. She would destroy everything under their nose, and let them come within 3 or 4 miles when she would run them out of sight in a few hours and commence destruction in another place. It would take the whole English fleet to guard the coasts of Great Britain against her & she could annihilate their commerce.[37]

The report of the *Wampanoag's* speed trial drew a mixture of consternation and incredulity, both at home and abroad. The *Army and Navy Journal* had already discounted any report which would come from naval engineers under the thumb of the "tyrannical" Engineer in Chief, and when the 17.75-knot speed was announced, Church greeted the news with indifference, remarking that the ship should have gone that fast in any case, and that the excessive weight of the machinery in producing such speed was the real issue. Conveniently forgetting the performance of Ericsson's engines in an identical hull, Church declared that the *Wampanoag,* regardless of her speed, was a failure because Isherwood's machinery took up too much room, leaving no space for crew, supplies, or armament. The relation of cause to effect, he reminded his readers, should always be kept in mind in viewing the execrable means used by Isherwood to achieve such an admirable end.

In Congress the *Wampanoag* immediately came in for censure because the accommodations for officers in the vessel were no larger than those on the ninety-day gunboats. Moreover, as Representative Elihu B. Washburne remarked, the cruiser undoubtedly was "the most extravagant ship to keep in commission in the entire navy" because she burned so much coal.[38]

Convinced that Isherwood had pulled "a 'smart' engineering trick," Ericsson believed that he had deliberately avoided a race with the *Madawaska,* because Ericsson's direct-acting engines had 60 per cent greater power than the geared engines on the *Wampa-*

*noag,* the inventor maintained. Isherwood, he related, had merely waited until the water was smooth and the wind fair, and then had allowed the *Wampanoag* to make a dash along the coast carrying a minimum of weight so that the vessel could attain a high speed.

Ericsson's loyal follower, William Church, improved upon the inventor's analysis by insisting that the "fresh breeze abaft the beam" had much to do with the *Wampanoag*'s speed.[39] In any event, he maintained, the cruiser was worthless because she was so cramped that after the trial sixty-five men had been removed to a receiving ship because of a lack of berthing space. Although the machinery took up such an inordinate amount of room, it had not developed the horsepower of armored British battleships, so it was obviously worthless, Church concluded. This vessel, like the other Isherwood "steam engine carriers" was just a white elephant, too lightly armed to be a formidable warship, too expensive to run, carrying insufficient coal for cruising, and useless while operating under sail. Consequently, he observed, "we could hardly wish a hostile power a worse fate than to be compelled to build and keep at sea a fleet of *Wampanoags.*"[40]

Across the Atlantic, naval and engineering authorities at first greeted the performance of the *Wampanoag* with extreme skepticism. *Engineering* published the full official report of the speed trials, March 27, 1868, because of the great interest in the trials shown in England; but the journal's editors saw little to commend in the designer of the machinery. The trial results, they admitted, were astonishing—so much so that until further details appeared, they declined to credit them, much as they did not wish to cast aspersions on the truthfulness of "our friends over the water."[41]

While *Engineering* displayed polite incredulity, the London *Engineer* minced no words in its opinion of Isherwood. Denying that the *Wampanoag* had developed more than a 16-knot speed or 3,500 horsepower, this journal maintained that Isherwood did not know how to design a good marine engine, and that England had nothing to fear from American competition so long as Isherwood remained as bureau Chief. It deplored only the fact that such a man and his designs had been foisted on such "an intelligent nation" as America, which was so keenly alive to the value of naval supremacy. "No chief constructor, no board of admiralty, no marine engineer in England," the *Engineer* proclaimed April 24, 1868, "has ever manifested the official incompetency displayed by the Bureau of Steam Engineering in the United States."[42]

While the British engineering journals blustered loudly about Isherwood's blunders, other authorities in England viewed the *Wampanoag* with a different attitude, quietly remarking, "We may well ask, what would *we* not give to have a class of vessels with such a heavy armament, and anything like the speed?"[43]

The British Navy had wasted little time pondering such a question. As early as 1866, before the *Wampanoag* had been completed, the Chief Constructor of the Admiralty, Sir Edward J. Reed, had begun to design a vessel to deal with the class of American supercruisers. Reports had widely circulated abroad about the purpose and designed performance of these vessels, and the prospect of a fifteen-knot commerce destroyer rampaging over the lanes of British commerce was more than sufficient reason for the Admiralty to act.

Reed produced the *Inconstant,* an iron-hulled cruiser utilizing full sail power as well as steam for the purpose of hunting down commerce destroyers. With heavy armament and a large carrying capacity, this vessel was considerably larger than the *Wampanoag,* displacing 5,780 tons, with her hull sheathed with copper on wood to resist fouling. With a length almost identical to that of the *Wampanoag,* the *Inconstant* had 5 feet more breadth and drew 4 feet more than Isherwood's cruiser. Although launched on November 12, 1868, the *Inconstant*'s speed trials were not held until almost a year later, at which time she achieved 16½ knots from a developed horsepower of 7,360.[44] Her small initial stability forced the use of 300 tons of ballast, which reduced her speed by one knot. Like her American counterpart, she was at once attacked for her excessive cost of operation, since the high speed she developed required a necessarily large amount of coal.

The widespread criticism which followed so closely upon the heels of the *Wampanoag*'s performance prompted Gideon Welles to obtain the opinion of three line officers on the merits of the vessel. On April 13, 1868 he ordered Commodores Melancthon Smith, Thornton Jenkins, and James Alden to Hampton Roads, Virginia, where they were to board the *Wampanoag* and accompany the vessel on another speed trial. This time the cruiser would steam from Hampton Roads back to New York, and would move under full steam or full sail, or a combination of both as the board of line officers saw fit. The purpose of their trip, Welles stated, was to see if the *Wampanoag* "comes up to a reasonable standard of efficiency, in view of the end sought in her construction."[45]

Reporting to the cruiser, the officers found that she had coal sufficient only for a one-day, full-speed trip to New York, and lacked a sufficient deck crew for carrying full sail. Despite these shortcomings, the vessel departed Hampton Roads on the afternoon of April 15, and gradually worked up to full steam, making thirteen nautical miles in sixty-six minutes against a strong wind and current. Almost as soon as the vessel was up to full power, she ran into "thick weather" which forced a reduction of speed to guard against collision. On the following morning, the *Wampanoag* briefly increased speed to sixteen and one-fourth knots, but dense fog closed in, forcing another reduction of speed. Afterward, there was no further opportunity for a speed trial, and the cruiser slowly steamed towards New York in the fog, arriving in the evening of April 16.

Commodores Smith and Jenkins, in a majority report to Welles, dated April 21, 1868, stated that despite the inability of the *Wampanoag* to complete a fair trial, they were satisfied that she could develop and maintain such speed that no vessel afloat could escape her or overtake her. Carrying a battery ample for her purpose, this cruiser, in their opinion, had satisfied the requirements for a warship of the greatest attainable speed, with sufficient armament for commerce destroying and self-defense.[46]

Since the "main and special purpose" for the *Wampanoag* no longer existed, the two officers suggested that the cruiser be modified in order to be more practicable as a peacetime warship. They agreed with Captain Nicholson, who had recommended in his trial report that a light spar deck be added to give more room for the crew. They also recommended that tests be made to determine whether some of the boilers and funnels might be eliminated. Those funnels remaining were to be made telescopic, so they would not interfere with the sailing qualities of the vessel.

The third member of the board, James Alden, heartily agreed with his colleagues that the *Wampanoag* should be altered to serve peacetime requirements, but he insisted on writing his own report in order to emphasize the inadequacies which he saw in the vessel. Comparing her machinery unfavorably with that of British ironclads, he asserted that Isherwood's engines took up far too much room for the horsepower produced, the Martin boilers were obsolete, the funnels seriously interfered with the vessel's sailing qualities, there was no room for coal and provisions, and the four-bladed propeller created a drag on the vessel when under sail alone. He recommended the substitution of a two-bladed propeller which could be detached and hoisted when not in use, the addition of a spar deck, and a rearrangement or removal of Isherwood's machinery, in order to make the vessel suitable for naval service. Despite these serious structural defects which he had criticized, Alden had to agree with the others that the *Wampanoag* was the fastest ocean steamer afloat; and as such, had undeniably attained the end originally sought by the department.

Before he was able to reply to the line officers, Isherwood was chagrined to see his cruiser taken out of commission on May 5. Becoming indignant at Welles's apparent agreement with the criticisms of the Commodores, Isherwood addressed a letter to the Secretary on May 15. In examining the report of these line officers, Isherwood said, he had become greatly annoyed by their inability to recognize that the *Wampanoag*'s qualities were just as essential in peacetime as during a war. "Modern warfare," he insisted, "includes much more than a yard-arm to yard-arm fight in mid ocean." The age of fighting sail was dead, no matter how strenuously these line officers tried to revive it.

Believing that his critics had paid mere lip service to the purpose for which the *Wampanoag* had been designed and constructed, Isherwood set down in detail, in a report to Welles, May 15, 1868, the reasons why this vessel had to develop tremendous speed. To compromise her ability to steam at such speed would destroy her entire reason for existence, he began. The machinery was large because the extra power required for each extra knot of speed was immense. Four funnels were necessary both to provide sufficient draft for the many boilers and to prevent too drastic a reduction of power if one or more of the funnels were hit during battle. In any event, he added, they did not interfere with sailing qualities because he had carefully located them so that the yards could swing freely and still not touch the funnels.

The criticism by Alden particularly incensed Isherwood. In the minority report of this line officer, Isherwood discerned motives beyond those of honest, if unenlightened, disagreement. Alden had brought up unfair and inappropriate comparisons between the *Wampanoag* and British warships, and he appeared to be criticizing the cruiser out of spite, intent only on emasculating Isherwood's masterpiece. The Engineer in Chief noted that Alden's criticism had been duplicated in the *Army and Navy Journal,* which Isherwood described in the same report to Welles as being "a weekly newspaper with a very limited circulation, devoted, as is well known, principally to the advocacy of a board of survey for the Navy, and with unparalleled malignity and falsehood to the professional abuse of the Bureau of Steam Engineering."[47]

Alden, like many another line officer, wanted to exclude professional technicians from influence in naval affairs. To achieve this goal, Isherwood explained, he had remorselessly attacked the work of the naval engineers, especially that of the Engineer in Chief. For this reason, Isherwood insisted that Alden was unqualified to be a critic, since he had this professional axe to grind.

Not surprisingly, the *Army and Navy Journal* sprang to the defense of Alden. The Engineer in Chief, Church asserted in the August 15, 1868 issue, had displayed his true colors by complaining about Alden's use of British ironclads for a comparison with the *Wampanoag*. Not only had Isherwood disputed this perfectly valid comparison, but he had had the gall to make an "outrageously deceitful" comparison of his own, in attempting to prove that the speed of the *Wampanoag* was the result of its machinery rather than its hull design. By attacking Alden at all, Church insisted, Isherwood had acted disreputably. Alden was a superior officer who should be free from the abuse of a subordinate staff officer. While this brave line officer had been risking his life during the war, Isherwood had been "conducting less dangerous battles with Martin's boiler. . . ."[48] The *Wampanoag,* Church reiterated, was an utter failure, and the *Army and Navy Journal* would continue to demonstrate this fact despite all the Engineer in Chief might do to cover his own errors by abusing his betters.

While the critics on both sides of the ocean were noisily depreciating the *Wampanoag* and Isherwood, the second of his cruisers, the *Ammonoosuc,* had her speed trials. With the same geared engines and a hull virtually identical to the *Wampanoag*'s, the *Ammonoosuc* would prove whether the speed of the first cruiser had been mere accident.

On June 15, 1868, the *Ammonoosuc* left New York Harbor for Boston, under orders to make a full-speed trial. Although fog prevented the vessel from attaining her maximum rate for more than the first and last hours of her trip, this cruiser gave a sufficient indication of her capabilities. Aided by smooth water on her maximum-speed runs, she traveled from Sandy Hook to Fire Island, a distance of 30 miles, in 1 hour and 47 minutes, averaging 16.8 knots. Approaching Boston, she broke out of the fog and briefly ran at full throttle, this time covering 50 miles in under 3 hours, averaging 17.1 knots. On both runs she carried only 32 pounds of steam pressure in her boilers, and for the entire trip, much of which was at reduced speed, she burned coal at only 2.65 pounds per horsepower hour. With no mechanical troubles at all, this steamer could have developed "a much higher rate of speed" with different coal, reported her chief engineer, since her engines had developed less power than the identical ones on the *Wampanoag*.[49]

Faced with the results of the *Ammonoosuc*'s trial, William Church muttered that Isherwood's "monuments of engineering folly" would have gone much faster with orthodox, proven machinery. At this, the Engineer in Chief could only shrug his shoulders and smile.

The remarkable performances of the *Wampanoag* and the *Ammonoosuc* have produced, in later days, exaggerated claims of their significance. Some writers have maintained that these vessels were responsible for keeping the British and French from intervening on the side of the Confederacy. Although the cruisers were intended to harass European powers in case of intervention and subsequent war, there is no indication that these vessels had influence on the determination of Great Britain and France to remain officially neutral. In any event, none of the cruisers were completed until after the conclusion of the Civil War, and the Isherwood-designed vessels were the last, as well as the best, of these warships.

The *Wampanoag*, by her demonstration of great offensive capabilities, has inspired others to give her credit for forcing a dramatic about-face in the conduct of British foreign relations with the United States. The threat of commerce destroyers, some have implied, was enough to make the British compromise and decide to settle the *Alabama* claims controversy which had exacerbated Anglo-American relations after the Civil War. Again, there is no evidence that the American cruisers played a significant part in the formulation of British policy, and the belief that this is one of the great unwritten facts of history must remain a pleasant fantasy.

Ironically, the *Wampanoag* had her most far-reaching effect on the British Navy. The very existence of such a vessel had produced a prompt reaction in the Admiralty. As Sir Edward Reed explained:

> There is one consideration superior to, and which always must and will outweigh the abstract thoughts of any of us, and that is the state of foreign navies; and I say that the *Inconstant* was designed—of her size, of her cost, of her length, of her horsepower, and with her armament and speed—expressly to compete with (and worthily to compete with) a very powerful class of American vessels, which was then under course of construction.... We could not afford, in this country, to wait and see whether they were to be successful or not; we had to produce a ship which would compete with them; and I say that fact affords the very best possible reason that you could have for laying down any ship—namely, the existence of ships in other navies that we wanted to compete with and be superior to.[50]

The British naval expert, Sir Thomas Brassey, amplified Reed's statement in a letter to the London *Times,* December 4, 1875, by declaring that when any foreign naval power even seriously contemplated, let alone produced a seventeen-knot vessel for the purpose of harassing British commerce, it was the duty of the Admiralty to answer this threat with equal if not superior vessels. For this reason, when the *Wampanoag* appeared, a new class of naval vessel became a necessity for the British Navy. Unwittingly, Benjamin Isherwood had become a leading contributor to British naval progress, although he may have been a prophet without honor in his own country.

In later years, Isherwood would receive the unstinted admiration and praise of naval officers and engineers, but in 1868 the achievements of the *Wampanoag* fell unheeded on a Navy brimming over with hostility towards its Engineer in Chief. He could take his full measure of satisfaction with the cruiser, but there were few to participate in his pleasure and to enjoy the vindication of his theories and designs. At a time when he should have been basking in the sunlight of his success, Benjamin Isherwood was to engage in a battle for his own survival.

# X. Line against Staff

As Engineer in Chief, Benjamin Isherwood supervised men as well as machines. Hailed as a leader who was "ever watchful over the welfare of his corps and a fearless defender of its rights,"[1] he took as uncompromising a stand for his men as for his engines. In a period when the status of naval engineers was in constant flux, he engaged in controversy as willingly and as belligerently as he had with his professional opponents over the merits of engineering theories and designs. In this instance, however, the opponents he took on were too many and too powerful. His struggles for his corps coalesced the many sources of resentment and opposition to him, and led inevitably to his downfall.

Upon taking office in 1861, Isherwood did not hesitate to involve himself in disputes on every level of naval administration. After joining with Senator Grimes to thwart a congressional attempt to encourage civilian appointments to the position of engineer in chief, Isherwood played a major role in initiating and guiding the naval reorganization proposals which culminated in the creation of his Bureau of Steam Engineering, in July, 1862.[2]

At the same time, he decided that the proper utilization of the naval squadrons blockading the Confederacy required the presence of a supervising engineer who would correspond to the commanding line officer. In early 1862, he recommended to Welles that the new post of fleet engineer be created to provide this function. Candidates would be restricted to chief engineers in the Navy, and they would qualify for this duty by special examination.

Chief engineers appreciated the opportunity for further advancement, but bitterly opposed the requirement of an examination. Regarding their rank as an achievement which marked final mastery of their profession, they argued that any further examination would only be an insult to their abilities and would degrade their present status. Fearing that this issue would annihilate both harmony and loyalty within the Engineer Corps, Isherwood's friends soon persuaded him to drop this requirement.

Despite the protests of line officers that this new position was unnecessary and would only feed the vanity of already presumptuous staff officers, the fleet engineer proved to be a busy and responsible individual during blockading operations of the Civil War. "To superintend and supervise everything in the engineer department of the Squadron, including the provision of its enormous material in proper quantities at proper times in proper places, was to make the success of the blockade mainly dependent upon him," Isherwood assured Welles, "for the ablest Commander could do nothing with unserviceable vessels, and with crippled or inefficient machinery every vessel was absolutely useless."[3]

The fleet engineer could keep vessels on station by restraining their departure for unnecessary repairs and by making certain that they were properly maintained so that breakdowns would be kept at a minimum. Without their steam engines in proper order, vessels could not operate effectively, since their sails, Isherwood continued to Welles, "were mere adjuncts, and it would have been better to have entirely removed them."[4] True as this was, line officers did not accept either the situation or the fleet engineer gracefully. They had resisted the attempt to create this higher position for engineers, and when it succeeded, their resentment turned against the man who had been primarily responsible.

Despite his own lack of formal engineering education, Isherwood recognized the need of such training for naval engineers. Moreover, it was not sufficient, he believed, that engineers alone should have a knowledge of machinery. If the age of steam power was now to take command, it seemed reasonable that all officers should be able to perform their duties with an appreciation of the capabilities and limitations of this motive power.

Welles agreed. His annual report for 1863 proposed that engineering education become part of the curriculum at the Naval Academy; if not for all officer candidates, at least for those who later would become engineer officers.[5] This program became a law on July 4, 1864, when President Lincoln signed a congressional act which established a course of engineering instruction for cadet engineers at the Naval Academy.

Although this was a step in the right direction, it was insufficient, Isherwood decided, that cadet engineers alone should receive a full course in this specialty. In his annual report to Welles the following year, he requested that all officer candidates be educated in practical steam engineering, since the Navy had become "almost exclusively a steam navy," and would inevitably continue to remain so. Just as naval officers centuries before had been unable to operate sailing ships, but served only as soldiers and gunners, the present naval officer was "a mere passenger on board a steamer," Isherwood argued. Under the current system, a line officer had deck duties which required little experience and which were of little importance compared to those of an engineer. Moreover, Isherwood tactlessly asserted, the line officer's duties required less ability for their discharge.[6]

Impressed with this argument, the Secretary in his own report requested that all midshipmen at the Academy receive education in steam engineering. With this knowledge a property of all naval officers, he suggested, there would be more efficiency in vessel operation and a possible saving in manpower. Naval officers, of course, would not receive instruction in engine design or naval architecture, which were true professional specialties, but would be taught only the necessary techniques of "engine driving."[7]

In November, 1864, Isherwood went to the Naval Academy at its wartime home in Newport, Rhode Island, and examined the graduating class of thirty-five midshipmen in the engineering knowledge they had so far acquired. After they demonstrated their proficiency by operating the screw gunboat *Marblehead,* Isherwood reported to the Secretary of the Navy that they had done well, considering the brief extent of their training, and had shown as much knowledge and skill as did third assistant engineers when first coming into the Navy.

The Naval Academy, however, did not immediately begin to produce full-fledged naval engineers. After moving back to Annapolis, it established a Department of Steam Engineering, which instructed all midshipmen in the rudiments of practical machinery operation. A steam building was erected in 1866, and before its completion, an Isherwood engine had already been installed to provide a practical model for the students. Although the Engineer in Chief was not able to produce a full program of engineering training for all midshipmen, he at least had the satisfaction of seeing his profession raised to the level of an academic discipline, necessary for the proper instruction of all naval officers.

In the years following the war, Isherwood labored to improve the pay and status of engineers already in the service. In February, 1865, he submitted to Welles a plan for the reorganization of the Engineer Corps in the Navy. He proposed that the old ranks be abolished and be replaced by an organization consisting of:

An Engineer in Chief

Twenty "Engineer Directors" (with duties similar to those of the present Fleet Engineers)

Chief Engineers (enough for one in each first and second class steamer)

"Engineers" (two for each first class steamer and one for each second and third class steamer, with duties similar to those of present First Assistant Engineers)

"Sub-Engineers" (two in each first class steamer, three in each second class, two in each third class, and one in each fourth class steamer, with duties similar to those of present Second Assistant Engineers)

"Cadet Engineers" (three in each first, second, third, and fourth class steamer, with duties similar to those of present Third Assistant Engineers)[8]

Isherwood also asked for an immediate increase in rank and pay for his men. The great amount of money being spent on machinery, he reminded Welles, in a report of October 12, 1865, would be worthless unless skilled operators of machinery were available. Civilian work was luring the best engineering talent away from the Navy, "leaving only mediocrity . . . for the government." Since talent would inevitably fall to the level of present pay and position, he urged a large salary increase for chief engineers "so that the prize in this lottery will be of sufficient value to induce first-class abilities to continue through the drudgery and small pay of the lower grades, in order finally to attain it."[9]

In the following year, Isherwood strongly endorsed a memorial sent to Congress by naval engineers requesting better pay and position, and he proposed to Welles, in a letter dated May 23, 1866, that the Navy should arrange to recruit professionally trained engineers, "graduates of the finest scientific schools in the country," in order to improve the caliber of the corps. The Navy should not have to rely on private designers and builders for its work, he once more maintained, since civilians were only interested in profits and would use the cheapest material possible. Design of naval vessels and construction of machinery would be safe only in the hands of the Navy itself, so he urged the growth and improvement of his corps, along with the development of government facilities.[10]

His reorganization proposal and requests for increased rank and pay came at an unfortunate time, since the end of the Civil War brought on a severe reduction in naval personnel as well as stringent economizing. Nevertheless, Isherwood's efforts had not been entirely wasted. During the war, Welles had regraded the relative rank of engineers so that, after March 13, 1863, they were on a more equitable level with line officers. As Chief of the Bureau of Steam Engineering, Isherwood now ranked with commodores, while his fleet engineers ranked with captains. Since staff officers previously had held relative rank only as high as commander, this was a considerable rise in status for the engineers.[11]

Isherwood's proposed reorganization, in 1865, failed to be adopted, but his first and second assistant engineers received a significant improvement in status the following year. By a congressional act of July 25, 1866, these engineers, for the first time, held commissions as officers in the Navy. Previously ranked *with,* rather than *as* commissioned officers, they had been constrained to the level of warrant officer on their ship, with all the frustrating inconveniences this subordinate position held. Although they now enjoyed the status of commissioned officer, they had to wait for some of the perquisites. A Navy Department order of the following year illustrated their problem by stating that, although they were no longer "steerage officers," these engineers could not immediately move into officers' quarters, since there was still a lack of wardroom accommodations because of the crowded conditions on board naval vessels.[12]

Whether Isherwood succeeded or failed in his efforts to improve the position of naval engineers made little difference in the attitude of line officers toward him. By the late 1860's their hostility toward the Engineer in Chief had grown to a formidable degree. To understand this attitude, one must examine the relationships between engineers and line officers during the Civil War.

Until 1861, there were relatively few steam warships in the American Navy, and the number of engineers was correspondingly small. Suddenly the wartime Navy became a steam Navy, and the line officers who had flourished in the days of sail now found their prestige and value challenged by "greasy mechanics" whose frequent lack of education and gentlemanly deportment could not disguise their invaluable function. A captain might still possess absolute authority on deck, but no longer could he depend only on the forces of nature for his propulsion; now he had to observe the requirements and limitations of his machinery; and all too often, it was the engineer who informed him what he could or could not achieve with his vessel. Machinery breakdowns were also inevitable, and in the heat of battle or during the tedious months of blockade duty, it was the engineer who bore the wrath of his commander when the vessel failed to respond properly.

Engineers, aware of their increasing importance, resented being treated as inferiors on their own warships. At times they were denied the officers' quarters to which they were entitled, and often they were forced to give up their berths to newly arrived line officers, regardless of rank. They even served as firemen and coal heavers, when line officers demanded that the engine-room crew stand deck watches and help to hoist sails. Slighted by the deck officers, engineers often believed the degrading treatment was calculated and sprang from a jealousy of line officers toward those who they felt to be interlopers.[13]

Engineers were not alone in their conflict with the line. All staff officers had experienced the frustrations of their inferior status, and naval constructors, paymasters, and surgeons joined the engineers in continual attempts to better their lot. Their common ambition was to be treated as equals, both within the service and without, while performing their own specialized functions and not intruding on those of the line. To this end they made a pointed distinction between *rank* and *command,* insisting that the former could be granted without the latter. What the staff officers did not want was command, either ashore or afloat, except of their own staff and over their own departmental functions. They recognized that line officers, trained to direct a ship in battle and to navigate her through all weather and waters, should properly hold command of the ship and should exercise absolute authority in their own sphere.

But along with equal rank, based on their date of commission as officers, the staff wished the honors, immunities, and pay of line officers. They asked for promotion based on seniority for professional grades, and for titles corresponding to their length of service and position. They requested that the duties of each grade be defined, as they were for line officers, and that within their own corps there be clear rules of precedence among the various grades.

They believed that staff officers with the highest rank and titles should not serve aboard ship, since they would outrank the senior line officer and yet be subordinate to his command. Nevertheless, staff officers of any rank, they argued, should not have to submit any longer to the petty indignities and restrictions which applied only to the most junior line officers on board a vessel.

Since the staff officer was often a career Navy man, he also resented being discriminated against by Congress, for it often appeared that only line officers could move up rapidly in the service. Even when a staff officer did advance, however, he could not enjoy the full benefits of his position, because a social stigma attached to his particular naval profession. A staff officer was forbidden to use the title of his rank in any address, whether official or social, so that an engineer with the rank of commander would be referred to only as chief engineer. In naval social circles, where rank and titles were of the greatest importance, this distinction neatly relegated all staff officers to an inferior status to the line, who so jealously guarded their monopoly of titles.

Welles's department order of March, 1863, was a great step forward for the staff, but, unfortunately, these officers were unable to consolidate their gains by legislative action which would insure the permanence of the higher rank conferred by the Secretary's order. Within a few years, this failure would cause them great regret.

At the same time, the line worked busily to restrain the advance of their naval inferiors. In April, 1864, a board of line officers requested that the department implement their plan which would assimilate the staff into the line, but would rank staff officers solely on the basis of their age, irrespective of length of service. By this plan, Isherwood, although engineer in chief and head of his own bureau, would rank with lieutenant commanders; and only one engineer in the Navy would rank as a full commander. Moreover, within each grade, where precedence would be based on length of service, the engineers would be near the bottom of the list, since their corps had been organized only in 1842 and they had less average service than line officers of the same age. Thus, Isherwood, forty-one years old at the end of 1863, had nineteen years and seven months of service, which was three years less than most line lieutenant commanders.

Such maneuvers by the line produced a quick response from the staff. Asserting that "the line are traditional, and one-idead; the tradition the past, the one idea themselves," the staff proclaimed to the world that it was being persecuted by these blind devotees of a departed age.

Everywhere the Line "spit upon our gabardine." They talk against us, write against us, frame "Regulations" against us, influence legislation against us, and strive (sometimes, as we have seen, with success) to incline the Department against us. Regarding one corps as having rights, and as being the "Navy proper," they would gladly drive out the rest as worse than encumbrances, whom they speak of at best only with patronizing condescension.[14]

After the Civil War this dispute became more severe. No longer was there any need for co-operation between line and staff, for the Navy had only minimal duties and responsibilities. In addition, the drive for economy had resulted in a partial return to the prewar days of sail. Engineers were no longer so essential, and the continuing activities of their Chief appeared senseless to many line officers. The intraservice fires which had been damped during the crisis of the early 1860's now broke out in full flame, fed by the boredom and frustration of a peacetime naval life.

With no battles to fight or torpedoes to damn, the line officers missed the adulation of a public which now had little need for naval heroes. Restless from their loss of glamour and prestige, the line officers turned against the men who were attempting to lift themselves to equality within the service. As much as they might depreciate the staff, the line found that the staff had its uses. Employing the staff, and particularly the engineers, as a "mudsill" for their own self-gratification, line officers could find some compensation for their loss of significance in peacetime through the lively pastime of berating these "glorified mechanics" for their temerity in striving to improve their lowly position. Without any doubt, the staff kept postwar naval life from being too humdrum, for, as one staff officer derisively noted:

The last grievance which the Line would wish removed is the offence of the Staff Officers; for with whom would they quarrel, unless with themselves, and what would relieve the dulness and insipidity of the Navy?[15]

The ideal solution, as Gideon Welles realized, was to amalgamate the engineers and the line entirely, so that line officers would become engineers as well as sailors and would command the engine room as well as the deck of their vessels. Realistically, the Secretary dismissed this utopian dream, since line officers were not only unalterably opposed to such a plan on principle, but in most cases, they were too old to receive engineering training or they lacked mechanical ability. Although he had suggested such a plan during the last several years, the Secretary knew that he would have to obtain considerable congressional support to push through a proposal which was as distasteful to the engineers as to the deck officers. Opposition within the service and the indifference of Congress precluded any serious attempt in the 1860's.[16]

Line officers soon discarded mere passive resistance to the aggrandizement of the staff. The only way they could be secure was to obtain control of the Navy itself, for in command of the administrative functions as well as the operations of the Navy, they could rescue their slipping prestige and consolidate their power.

The decline of line-officer control of the Navy dated from 1842. At that time the administrative Board of Navy Commissioners, comprised of line officers, had been replaced by the bureau system. By the mid-1860's there were eight of these semiautonomous offices loosely grouped under the Secretary of the Navy. A lack of close professional supervision by the Secretary tended to produce overspecialization, duplication, and waste, as the bureau chiefs, absorbed in their own limited branches of naval service, rarely co-ordinated their operations with those of the other bureaus.

Several of the bureau chiefs were staff men, and their power and prestige had grown tremendously with the naval expansion during the Civil War. Benjamin Isherwood and John Lenthall particularly benefited from this expansion, and their control over the determination of policy as well as administrative decisions infuriated line officers who were tied down by their battle-field responsibilities and were unable to participate effectively in department activities.

Official business between Congress and the Navy Department went through the Secretary's office, but often the most important decisions were formulated through direct personal contact between bureau chiefs and congressmen. The staff had particular advantages over the line whose officers were serving on distant stations, while the senior staff officers tended to cluster in Washington, where the bureaus spent most of the Navy's money. Their contacts with Congress were particularly effective, since the channeling of Navy funds into construction, repair, and supplies provided an effective tool to establish "a definite community of interest" with representatives of seaboard states.[17] Such power of men like Isherwood and Lenthall to allocate funds appropriated by Congress proved particularly galling to line officers who lacked both opportunity and means to exercise commensurate influence. While the staff grew inexorably in power and size, the line desperately sought a way to break the source of its strength.

By 1865 the line had determined to act. In the debates over the naval appropriation bill early in that year, a group of congressmen, led by Henry Winter Davis, proposed a reorganization of the Navy Department in order to put an end to the weakness of the present structure. The new department would feature a Board of Admiralty, comprised of several senior line officers, which would advise the Secretary of the Navy and generally supervise both the operations and administration of the department.

Welles admitted that such a group would relieve his office of much detail, if it were properly set up and regulated. The possibility of this, however, seemed remote to him. The Board of Admiralty scheme was little more than "a favorite theme ... to give naval ascendancy in court sessions," the Secretary noted. The board would inevitably break up into jealous factions, and despite its "advisory" purpose, it would doubtless end up dictating to the Secretary of the Navy. "It would not be beneficial to the government and country," Welles concluded.[18]

Congress appeared to agree. Representative Alexander H. Rice vigorously attacked Davis's proposal in the House, and Senator Benjamin Wade obtained only two votes in the Senate to support the scheme. However, this would not be the last of the matter, Welles realized, for too much was involved. Such congressmen as Davis, Thaddeus Stevens, and House Speaker Schuyler Colfax had been assisted by senior naval officers including Samuel du Pont and Charles Wilkes; and, as Secretary Welles noted, in both the Senate and the House, they were able to launch a "deliberate and mendacious assault" on the Navy Department while promoting their plan. The Board of Admiralty idea might have merit, Welles admitted, but not when backed by "a few party aspirants in Congress, and a few old and discomfited naval officers, with some quiddical [sic] lawyer inventors, schemers and contractors."[19]

Within three years another and more serious assault on the department was in full swing. In December, 1867, Senator James Nye, of Nevada, proposed the creation of a "Board of Survey" made up of three line officers, all with flag officer rank. This appointive group was to co-ordinate the operations of the department which was in deplorable condition, William Church asserted, because of the "lack of harmonious action" among the various bureaus. "The principal source of trouble," he explained in the *Army and Navy Journal,* December 21, 1867, "is the Bureau of Steam Engineering; and Mr. Isherwood is the Jonah who has brought such bad fortune to the whole ship's company." The remedy, he implied, was obvious; throw the Jonah overboard.[20]

The lines of attack on the bureau system now became apparent. Since 1865, although in a period of drastic economizing and tight restrictions on steaming, the Engineer in Chief had continued his costly experimentation and had built dozens of engines in apparent indifference to the problems of the rest of the naval establishment. He had been able to do this, his opponents believed, only because the bureau system had afforded him such independence of action and unchecked power of decision. During the war, his prodigious efforts might have been essential to naval success, but now they were no longer vital. Isherwood's power and independence now symbolized the waste and irresponsibility plaguing a postwar Navy no longer able to afford the massive expenditures and far-flung programs of previous years. To tear down Isherwood was to destroy the bureau system, for to many observers he had become the logical and deplorable end result of the present naval administrative structure.

Isherwood thus became a whipping boy to advance the cause of the line and its Board of Admiralty. Leading the efforts to publicize and promote the Board of Admiralty plan, the *Army and Navy Journal* now found another reason to attack the "agrarian Chief who has hitherto monopolized all the steam department of the Navy, to the exclusion of everyone else." Isherwood, Church asserted, had made persistent efforts to destroy the Navy's efficiency, and the entire decline of the service since the end of the war could be laid at the feet of this irresponsible bureau Chief who was protected by the chaotic bureau system.[21]

The fact that the Engineer in Chief had power and independence appeared to be sufficient reason for many to assume that he was corrupt. The Board of Survey, one line officer declared, was necessary to "break up the steam ring," to check the "profuse expenditures" of the bureau Chief, and to restrain him from doctoring the reports of steam trials on his machinery. The board would, of course, be made up of line officers, because only these men could be trusted. "We have never yet heard of their retiring from a bureau with an ample fortune," remarked one officer who had apparently assumed that Isherwood's large fortune could have come only from corrupt transactions within his bureau.[22]

Only admirals could be trusted with the fortunes of the Navy, many officers claimed.

> The presumptuous aspirations of the Staff will never be checked till we have a board of Survey consisting of admirals alone, who, knowing what the service requires, could guide the Secretary of the Navy, put an end to "Wampanoags," mismanagement, and extravagance, weed out the sources of discontent, and by a firm, just, and impartial government, restore harmony and efficiency to the Navy.[23]

Naturally, the honesty and fidelity of admirals was beyond question in this regard. "These officers," Church maintained, in the *Journal,* March 14, 1868, "are so elevated by rank and position above the conflicting opinions and desires of all branches and departments of the Navy, that they may be safely trusted to conduct its affairs with wise impartiality." Thus there was no reason for engineers, surgeons, and paymasters to resist the Board of Survey idea. It was not "a sort of flank movement of the Line on the Staff," he explained, but rather a heroic attempt to cleanse the Augean stables of the Navy Department by sweeping out men like Benjamin Isherwood.[24]

Constant suasion and pressure by the line produced influential congressional support. Senator Grimes now decided that there were too many engineers in the Navy and that the best thing for the service would be a thorough reorganization of the Engineer Corps. "He is intensely hostile to Isherwood . . . ," Welles observed in his diary; and he went on to note that Grimes had "imbibed all the prejudices of certain line officers against the engineers, who are becoming a formidable power and rivals with the line officers in the service."[25]

In Congress, Senator Nye declared that "there is a war going on between the engineering department and those who belong to the

Navy proper...." He urged the creation of a Board of Survey because of the serious division extending throughout the Navy "between the sailing portion, the professional portion of the Navy, and the engineers." He insisted that the department had been unforgivably "clannish" in relying on Isherwood instead of using civilian engineers for their work.[26] Such pronouncements of the "party line" of the sailing officers merely disgusted the Secretary of the Navy, who scornfully noted in his diary that "blundering, plundering Nye, without honesty or integrity, but who has some pretensions to coarse humor, got in a fog and bellowed about the engineers and their rivalry with the officers."[27]

Welles might dismiss the crude and inept attacks of men such as Nye, but there were few so knowledgeable as the Secretary. The concerted assault on Isherwood in Congress and by the press slowly made its impression, and even the breath-taking accomplishments of the *Wampanoag* were easily brushed aside by those determined to destroy Isherwood and his corps in order to achieve a reformation of the Navy which would result in line control.

During and after the *Wampanoag*'s trials, Isherwood observed the campaign of the line with distaste. Although he had been preoccupied with his own work, he had not been totally oblivious to the political maneuverings about him. Located in Washington where he could be aware of all that occurred in the Navy Department, he soon realized that the opposition to him by the line was a co-ordinated, well-organized movement. "They have a delegation here in Washington, among whom are Alden, Worden & Rogers," he reported gloomily to Fox in a letter dated March 5, 1868, "and they have an executive Committee in New York. A regular assessment—monthly—is levied on the line officers for the expenses of the campaign."[28]

It was no surprise to Isherwood to discover that the line officers had "bought up" with their campaign fund the *Army and Navy Journal*, and, in addition, the services of Isaac Newton. This engineer, a friend of Ericsson and Stimers, and associated both with the original *Monitor* and the light-drafts, had been hired "to write the engineering articles abusing me & these vessels," Isherwood stated. The line officers had also brought Newton to Washington "to enlighten the Naval Committees as to my engineering enormities," Isherwood wryly explained, and for his efforts Newton was supposed to be receiving $3,000 annually plus his expenses.

As a result of this thorough campaign, Isherwood explained, the Senate and House naval affairs committees were now deliberating whether he should be turned out of office for producing the *Wampanoag*—and "rather thinking I should." At the same time, Representative Robert T. Van Horn, of Missouri, had introduced a bill in the House intended to remove Isherwood from office and replace him with a civilian. In the face of this hostile action, Isherwood concluded, "I keep my valise packed in readiness, and as soon as the new Admiralty Board comes into action, I expect to be ordered to sea in one of my own gunboats."[29]

The Engineer in Chief was premature in his pessimism. To oust him from office demanded more than the resentment of line officers. Congress, and the Senate in particular, would have to be aroused to action, and this group of men was notoriously difficult to stampede. To bring about Isherwood's removal demanded the efforts of someone who could coalesce the fervent but disorganized opposition to the Engineer in Chief. The logical place to look for such a leader was within the Navy itself, for here emotions ran high and much was at stake not only for individuals but for the service as a whole.

Leading the fight for the line and, consequently, directing the opposition to Isherwood, was the flamboyant Civil War hero of the western waters, Vice Admiral David Dixon Porter.

Son of the famous David Porter who had commanded the *Essex* in the War of 1812, and half brother of David Farragut, David Dixon Porter was the essence of the line officer; a career Navy man who had been reared in the age of sail, and whose gallant, impetuous, and imaginative action in the Civil War won him money, public acclaim, and a swift rise in rank until, in 1865, he became the first vice admiral in the history of the United States Navy.

Porter's career during the Civil War was glamorous, and his personality matched his deeds. Clever, yet devil-may-care, he cheerfully discarded any semblance of tact while speaking his mind on men and events. Unlike the formal, studied style of Isherwood, Porter's writing was remarkably spontaneous for his era. His ebullience carried over into his actions as well, and, as the firebrand of the Navy, he caused as much exasperation as admiration in the department.

At the conclusion of the Civil War, Porter received an assignment as superintendent of the Naval Academy, which had just returned to Annapolis. Life at the academy proved dull after the excitement and dangers on the Mississippi and Red rivers, and Porter chafed at his inactivity. Despite his taking an active interest in the midshipmen and their curriculum, his attention soon strayed to Washington where maneuverings for power, both in the postwar Navy and in the political world as a whole, were enlivening the affairs of the capitol. Eager to partake in this political turbulence, Porter established and cultivated contacts with friends in the Navy Department and in Congress. If anything was going to happen in the Navy, David D. Porter made certain that he would participate.

Porter's major failing was his excessive enthusiasm. Hardly a reflective person, he proclaimed what he believed at any given moment, heedless of its degree of consistency with past statements. "From my personal intercourse with him," recalled Daniel Ammen, "I feel assured that he never wrote what he did not believe to be true; but what he believed he did not think worth while to inquire into, as a matter of fact." This "unhappy idiosyncrasy" would plague Porter throughout his career by continually forcing him into extreme positions from which he could gracefully withdraw only with difficulty.[30]

In 1866, Porter asserted to Chief Engineer W. W. W. Wood that "there is no class of officers in the Navy who are entitled to more commendation than the engineers. None in the Navy have to pass through so hard an ordeal, and in battle their position is one where the utmost coolness is required, and where there is much danger."[31] Yet within a year he was furiously excoriating both the engineers and their Chief, by crudely casting doubts on their bravery and patriotism, as the line-staff controversy reached its peak.

Similarly, his view of Isherwood changed radically during Porter's career. He had warmly commended the engineer after their Mexican War service together. Yet by 1862, Porter was complaining to Fox, in a letter dated March 28, that Ericsson should be made engineer in chief regardless of the efforts it might take to remove Isherwood. It was deplorable, he then stated, that a genius like Ericsson remained unrecognized while "charlatans" were "fattening on the crumbs from the public crib."[32] By 1868, when he was in the midst of his attack on the Chief of the Steam Engineering Bureau, Porter managed to recall that, in the Mexican War, he had removed Isherwood from the engine room for incapacity; and that

Isherwood had grown only in age since that incident, since he avowed to Representative E. B. Washburne, "his incapacity still sticks to him."[33]

Porter's opposition to Isherwood became a full-scale effort in late 1867, and continued for the next two years. While the Admiral clearly considered Isherwood to be both the leader and the symbol of the staff's challenge to line control, there were more than intraservice factors involved in the Admiral's opposition.

After the war and during Andrew Johnson's administration, the sharp division between Moderate and Radical Republicans drew partisans from the military services. Welles and much of his department aligned themselves with the Moderates, seeking a mild reconstruction policy through early admission of the seceded states and a speedy restoration of the Union. As a prewar Democrat, Welles soon rebelled against the policies urged by such men as Seward, Stanton, and Sumner; and the Navy Department, consequently, was marked as a prime target for those who planned to seize power in the elections of 1868. Isherwood, like Welles, had been and would always be a Democrat. Without playing an active role in politics, his sympathies nevertheless were apparent, and he supported Welles and the Moderates throughout the late 1860's.

Porter, on the other hand, threw in his lot with the Radicals. Craving power, the Admiral recognized that in the upheaval following the seemingly inevitable election of Grant in 1868, he could become a major if not the leading voice in the administration of the Navy. In this way, he could consolidate the power of the line and protect it from further encroachments, initiate a number of his own pet theories on naval operation and administration, and gratify his irrepressible desire to "run his own show."[34]

The most effective means to this end, he decided, would be to obtain congressional and public support by vilifying the present administration of the Navy Department, and especially the Engineer in Chief, who was the most vulnerable to attack. On December 9, 1867, he wrote to William Church, explaining that the Navy had "determined to put a stop" to Isherwood, and that Porter would lead this effort by making a number of literary contributions to the influential service journal.[35] At the same time, he wrote to the editor and proprietor of the Washington *Chronicle,* John N. Forney, asking for support for the Board of Survey bill, and enclosing an article on the Engineer in Chief which was designed to demonstrate the need for the board of line officers.

Although he informed the editor of the *Chronicle* that it was not his desire to abuse anyone, but rather "simply to state facts," Porter began his anonymous article by declaring that anyone reading Isherwood's latest report to Welles would "be struck with the depravity that pervades the document." Isherwood, he maintained, had plunged the country into debt by his wasteful and useless experimentation. As an example of such waste, the "secret trial" of the *Wampanoag* [presumably the first steaming at sea to test the engines] had cost the government over $500 a day.[36]

The engineers whom Porter had so handsomely complimented in 1866 now appeared to the Admiral in a different light. As a class, he wrote to Forney, December 19, 1867, they "are not loyal, that is they would not at present fight to hold on to what we have won, and they belong to that element which would put all the rebels in power again if they had the chance." Their leader, Isherwood, would gladly sacrifice all the rights of line officers, Porter believed; and, therefore, it was fortunate that Grant would soon be President, because he would set matters straight by heeding the line.[37]

Isherwood had nearly destroyed the Navy, Porter insisted, because he had been responsible for the light-draft monitors. These ironclads, the Admiral continued, not only were badly designed, but also "were superintended by *semi-rebel* engineers," who had guaranteed the failure of these greatly needed vessels. It was a shame, Porter concluded, that Isherwood had been in office during the war, because he had "opposed that great genius Ericsson" who, receiving support only from line officers, had no chance against the firmly entrenched bureau Chief.[38]

Directing his next article to the *Army and Navy Journal*, December 24, 1867, Porter reported that "all Congress is alive just now to the importance of killing off this little mischievous engineer, who is disgracing us all over the world, and killing the navy."[39] Congressman Frederick A. Pike, of Maine, chairman of the House Naval Affairs Committee at that time, had just spent a day with Porter to obtain information about the Engineer in Chief; and was now "fully alive to the importance of getting rid of Isherwood . . . ," Porter enthusiastically related. If Congress could not remove the engineer, then the Bureau of Steam Engineering would have to be abolished, Porter insisted; for "we must leave nothing undone to get this charlatan removed from the position he so unjustly holds."[40]

Convinced that no effort should be left untried, Porter contemplated investigating the sources of the bureau Chief's large personal fortune, hoping that evidence might emerge which would prove that Isherwood had made his money by the illegal use of his position. If this failed, Porter could always rely on the apparent extravagance of the steam bureau's postwar operations to impress an economy-minded Congress.

Throughout late 1867 and into 1868, Porter solicited support from the House of Representatives by alarming congressmen with accusations of Isherwood's malfeasance, while simultaneously promoting his panacea, the Board of Survey. Dozens of letters poured from the prolific pen of the Admiral as he labored to persuade these congressmen of the current deplorable state of the Navy—and who was responsible for it.

In the Senate, as well as in the House, Porter applied suasion to good effect. When Senator Nye introduced his Board of Survey Bill in 1867, Gideon Welles observed that Porter had obviously originated it. Unfortunately, Welles added, the Admiral was totally unfit for the administrative office he sought through such legislation. Although brave and resourceful as a commander, Porter was "uneasy, scheming, ambitious, wasteful in expenditure, partial and prejudiced as regards officers," and "a most unfit administrator of civil affairs." Welles was certain that there was no need for such a board as Porter proposed. During the war the Navy Department had functioned efficiently, he believed, even when it expanded tremendously in size; and now that it had become drastically reduced through postwar retrenchment, there was even less reason for reorganization. The whole idea was simply absurd, for the Board was only "a miserable contrivance" of Porter's "to get place and power in Washington."[41]

Nevertheless, Porter had convinced influential men of the necessity of his scheme. According to Gideon Welles, Senator Grimes, a staunch defender of the department during the war, had fallen under Porter's influence. Welles observed that the Admiral had established a clique at the Naval Academy, and one of its members was John G. Walker, a nephew of Senator Grimes. Under Porter's direction, Walker had journeyed often to Washington to visit his uncle, and soon the Senator had grown distant toward Welles, had changed his views on the administration of the department, and had begun to support ideas he once would have scorned. In poor health at this time, Grimes was particularly vulnerable to the "malign influence" of the Admiral, and Welles discerned that Porter was effectively utilizing Walker as his "unconscious dupe and tool" in order to sway the enfeebled Senator.[42]

The success of the *Wampanoag*'s trial gratified Welles, but he soon realized that it would not still the clamor of Porter's clique. "Isherwood has exerted himself wonderfully to make his engines a success and has been sustained by the Department in his efforts," Welles noted in his diary approvingly. With the opposition steadily mounting against him, Isherwood stood in great need of this vindication of his theories and actions as bureau Chief. "I am glad, on Isherwood's account as well as on my own and that of the service, of this favorable result," Welles remarked, relieved that their "doleful predictions of failures" had backfired on the line officers.[43]

Reviewing the activity of the Admiral, Welles concluded that Porter's intrigues against Isherwood and the staff were the discreditable actions of a man obsessed with position and power. "With some very good qualities as an officer, he has some great faults and is wholly unfitted for an administrative place here," the Secretary decided. "In his restless, suggestive nature, the Department would experience infinite evil. He should be kept afloat and in active service, but with a taut rein."[44]

Unfortunately, Welles was unable to restrain the Vice Admiral. Porter continued his relentless attack on Isherwood and sought for ways to compromise or ridicule the bureau Chief. Sometimes with burlesque, other times with vituperation, Porter filled the pages of the press with a steady flow of imaginative, clever, and all-too-effective abuse.

Faced with the problem of beating down the concerted opposition of the staff to his Board of Survey proposal, Porter attempted to drive a wedge between various branches of the naval service. Arguing that the engineers were mainly at fault, he tried to convince other staff officers that the Board of Survey would not harm them, but would only eliminate the rotten and corrupt elements of the Navy. Even the engineers themselves might be spared, he suggested, if their Chief were to go. "When I have Isherwood shorn of his glory and sent back to the tribes of Israel where he belongs," he wrote to naval Paymaster J. George Harris, February 7, 1868, "I may withdraw my evil intentions toward the Engineer Corps, notwithstanding they have set themselves up as the preeminently scientific and 'only educated gentlemen' in the Navy!"[45]

In making Isherwood his primary target, Porter was convinced that once the bureau Chief fell, the opposition of the staff to naval reformation would collapse. In any case, the removal of Isherwood would be a fine reward for Porter's efforts. "I assure you that I never get tired at pegging away at Isherwood . . . ," he wrote, October 31, 1868, to Isaac Newton, the former naval engineer who was busily waging his own war against the Engineer in Chief. Reviewing his strenuous campaign, the Admiral added, "After all the efforts made to dislodge him it would be hard if in the end we failed to bring him to grief."[46]

Porter once more was in his element. If there were no more mortar fleets to command, if there were no more gallant and dashing expeditions to lead against the rebels, at least there was a war of another kind, and a great victory to be won. The boring, frustrating life in a peacetime Navy would not do for Porter; he sought excitement and action, and by dealing with Isherwood and the staff, he had found it.

Isherwood did not help his own cause. Too busy with his duties to spend time in replying to the attacks of various critics, the Engineer in Chief permitted his personal support to dwindle. When he did make a public statement, however, matters became even worse. The line-staff controversy, and Porter's opposition to Isherwood in particular, undoubtedly flourished as a direct result of Isherwood's annual bureau report to Welles for 1867 which was published with the Secretary's report on December 2, 1867. With his characteristic bluntness and indifference to the consequences of his statements, Isherwood thrust a cool but uncompromising challenge in the face of his adversaries.

"The navy is now, and must ever continue to be, exclusively a steam navy," Isherwood began. Engineers, henceforth, would play a predominant part in developing the naval service. Moreover, the growth of scientific knowledge and the need to employ this knowledge effectively demanded greater skill and experience among naval engineers. To obtain such skilled engineers to operate the increasingly complicated and expensive equipment demanded in a modern Navy, the government had to be willing to give such men sufficient pay and prestige. Because it had so far refused to do so, the Engineer Corps had been seriously depleted of talent. In the first year after the war, Isherwood pointed out, 83 regular engineers—among whom were many of the most highly skilled technicians—comprising 31 per cent of the entire corps, had resigned from the service because of the bleak future it afforded them.

During the war, he continued, engineers had been invaluable to the cause of the Union. The South had failed to obtain skilled engineers, and this lack had proved fatal. "They had despised the mechanical arts and sciences," Isherwood explained, "and by those arts and sciences they fell."[47]

Now, Isherwood declared, "the war and the progress of the age have changed our naval tactics, naval ships, naval machinery, and naval organizations; they have swept away many of the mouldy prejudices of an effete regime." Only engineers could continue naval progress through new mechanical inventions and scientific discoveries, and these men, Isherwood implied, would stand as a bulwark against the forces of reaction who wished to recapture those "mouldy prejudices" and reinstate their "effete regime."[48]

The government should encourage and continue to finance the elaborate experiments which he had been conducting, the Engineer in Chief urged. And because millions of dollars would be needed, it was only common sense that the Navy should employ the best technicians and revise the pay and rank of naval engineers. Line officers had benefited by the addition of five new grades of rank since 1862, and the staff had not kept pace. Engineers had formerly ranked as high as commander, only one grade below the highest line rank of captain; but now, he pointed out, this was no longer the case. With the addition of the three flag grades of admiral, vice admiral, and rear admiral, the chief engineers, despite Welles's order of March, 1863, had actually suffered a practical reduction in rank. To remedy this situation, Isherwood proposed that chief engineers with five to ten years of service rank with commanders; ten to fifteen years, with captains; and over fifteen years, with commodores. The chief of the bureau, of course, would then rank with rear admirals.

Isherwood's report stirred up a hornet's nest. His demand for increased rank, especially the grade of rear-admiral for himself, threw line officers into a rage. That his argument was essentially reasonable was irrelevant; he had committed an unforgivable heresy and had to be punished.[49]

"Oh ye Gods and little fishes," David Porter exploded in a letter to the *Army and Navy Journal,* December 9, 1867, when he heard of the Engineer in Chief's proposals, "will the country ever learn wisdom, or will the government ever cease to place Isherwood at the head of steam bureaus!" Isherwood, he fumed, "verifies an old saying—'give a beggar a horse and he will ride to the devil.'" It was obvious to the Admiral that the bureau Chief was building his own empire, greedily grasping for power and prestige for himself and "the various satellites revolving around him." It was incredible, Porter marvelled, that anyone would take the Engineer in Chief seriously, for he had wasted millions in producing worthless engines, and this irresponsibility obviously could not continue much longer. "The country cannot stand Mr. Isherwood," Porter declared. "He is too costly an elephant."[50]

The Engineer in Chief's demand for higher rank and pay were an insult to the brave heroes of the Civil War, the Admiral maintained. After remaining comfortably at home while line officers freely offered their lives for their country, Isherwood now attempted to claim the fruits of their victories. Who was this man to have the gall to demand such favors? After all, Porter contemptuously recalled, "Never once has Mr. Isherwood heard the whistle of a hostile shot, or placed himself where he could strike a blow for his country." Isherwood's only contribution to the Civil War, Porter declared, had been to share with Jay Cooke and Company the dubious honor of fixing upon the nation its magnificent public debt. The Engineer in Chief must have thought this debt a blessing, the Admiral observed sarcastically, so ardently had he labored to increase it by his expenditures.[51]

To his friend Commodore Henry Wise, a senior line officer and chief of the Bureau of Ordnance, Porter wrote, "The article of Isherwoods was very stupid and almost as bad as his Steam Engines. Why don't you line officers in the Department kick him for his impertinence in claiming the Rank of Rear Admiral!" The Board of Survey was now more necessary than ever, the Admiral insisted. Although not as rigorously partisan to the line as Porter would have liked, Nye's bill had to be implemented as soon as possible in order to "keep the 'staff' in their proper places, check the impertinences of the civilians, and even make it a Naval Department, which it is not now. . . ." The line officers had to rescue the Navy from Isherwood and Welles, Porter realized, and he would leave no stone unturned to effect this outcome.[52]

Warning other staff officers not to rally to the side of the Engineer in Chief, Porter urged them to trust the line officers and not to resist, because the line had the power to destroy any opponent. "Isherwood has put in a claim to be made rear admiral," he notified Surgeon Ninian Pinkney, December 31, 1867, "and to punish him for his folly we intend not only to strip him and the engineers of all honors, but to make them the most inferior corps in the Navy. Now when I say this," he warned, "understand we have the power to do it. . . . We intend this winter to throw the [engineers] so clear overboard that there won't be enough of them left to tell the tale."[53]

By his strength of conviction, his tactlessness, his indifference to service politics, his contempt for outmoded traditions of naval service, and his devotion to technological progress, Isherwood had cast down the gauntlet to the most powerful group within the Navy. No longer was he fighting civilians, with whom he could deal in his own sphere of accomplishment. He now had antagonized the majority of officers in the Navy, both by leading the staff against the line and by exalting his corps and his own engineering achievements in the face of the forces of tradition and reaction.

There were few to whom he could turn for aid, even if he would do so. Caught in the grip of economic retrenchment, Congress was unsympathetic with his work and his program. Indifferent to technological progress in a period of postwar exhaustion, the Union was not interested in building a modern Navy at great cost. That Isherwood had designed and produced the fastest ship in the world meant little to a nation which had turned its face away from the sea. War with Europe no longer was imminent, and there was no more American merchant marine of any consequence to protect. Why should the nation have to support a large Navy with fast, but expensive cruisers?

Many groups, within the Navy and without, grew increasingly

impatient with the Engineer in Chief. The elections of 1868 would bring in an administration which would have little sympathy with Welles's conduct of the Navy Department. There would be opportunities for great profits and power in the department if there was an overthrow of the Welles regime, and there was no better way to insure this overthrow than to blacken the reputation of the regime so thoroughly that the pressures for reform would become irresistible. As the election approached, this fact became strikingly clear even to Benjamin Isherwood.

# XI. Triumph of the Reactionaries

In the midst of intraservice controversy and disputes with civilian engineers, Isherwood's bureau became involved in a petty business dispute which eventually widened into a serious political issue. On June 15, 1867, Isherwood sent his friend Theodore Zeller to New York city to examine certain used machine tools which were for sale and which might be useful for the manufacturing and repair operations in the Philadelphia Navy Yard, of which Zeller was chief engineer. These tools were being offered for sale by John Roach, the iron manufacturer and machinery builder who, in expanding his business, had recently consolidated his operations in the newly purchased Morgan Iron Works and was selling off his used equipment from the Etna Iron Works in order to replenish his capital and eliminate duplicate tools. Zeller went to the Etna Works to inspect the tools and determine whether they were in good shape and were offered at a reasonable price. If such was the case, Zeller was to submit a requisition to the Bureau of Steam Engineering for any of the tools he wished to buy.[1]

Preoccupied with the approaching engine tests of the *Wampanoag*, Zeller made a brief inspection of the tools while they were in operation, approved a number of them for purchase, and turned the matter over to an associate to work out the details. In all, some 21 tools, including a planing machine, drill presses, shaping machines, and double-headed lathes, were purchased from Roach for a total of $42,425. In the winter of 1867 and the spring of 1868 they were delivered to the Philadelphia Navy Yard, assembled there, and put into operation.

The purchase of tools from a New York firm for the Philadelphia Navy Yard brought immediate protests from Philadelphia tool manufacturers. Two of the leading firms, Bement and Dougherty, and William Sellers and Company, were particularly incensed, since they had assumed the Navy would order from them. Engaging the services of John Rowbotham, master machinist of the Philadelphia Navy Yard, these firms had previously approached Zeller with excessively high quotations for the price of new tools, relying on the abilities of Rowbotham to convince Zeller that their offers were fair. To their dismay, Isherwood suddenly had intruded and sent Zeller to New York to buy used tools from Roach, whom the Philadelphia firms considered an interloper in the tool manufacturing business.

Failing to obtain orders through normal channels, the two Philadelphia firms determined to seek political assistance. Contacting William D. Kelley, a congressman from Philadelphia and member of the House Naval Affairs Committee, they persuaded him to examine the tool purchases in the hope of discovering irregularities. Spurred to action by the approaching congressional elections, Kelley quickly introduced a resolution in the House calling for the investigation and arranged to be named as chairman of the investigating subcommittee.

In addition, Rowbotham began to attack Isherwood's bureau from another direction. Realizing that the line-staff controversy was in full flame, he enlisted the aid of a former employer, G. N. Tatham, in order to approach indirectly the commandant of the Philadelphia Navy Yard, Rear Admiral Thomas O. Selfridge, who presumably would be eager to aid any criticism of staff officers. Tatham's letter of June 25, accusing Isherwood's bureau of favoritism and corruption in the purchase of tools from Roach, appeared to Selfridge to have come from a disinterested observer, and the line officer quickly forwarded it to Secretary Welles.

Welles read the letter with consternation and sent an immediate inquiry to his Engineer in Chief, who was in New York at the time. "The charges contained in [the letter]," he informed Isherwood, "are of a serious nature and if untrue should be at once disproved." If they were true, he added, he wished an immediate and full explanation, since Tatham's statement was "of a very extraordinary character." It could not be brushed aside, because a copy of the letter had also gone to the House Naval Affairs Committee. In any case, Isherwood should return to Washington at once, advised the Secretary.[2]

After conducting his own inquiries, Isherwood replied to Welles that the purchased tools had been "absolutely needed" for repairing steamer machinery at the Philadelphia Navy Yard. Without them, the bureau would have had to contract with outside parties for the work, which would have meant high costs and delays. Moreover, the tools had to be ordered at once, since at the expiration of the July 1 fiscal year, all surplus funds not earmarked for purchases would have reverted to the treasury.[3]

Zeller, he added, had first visited the larger tool manufacturers in Philadelphia and obtained their prices before he made the selection of Roach's tools. Since the local prices had been between 40 per cent and 50 per cent higher than those in New York, Isherwood concluded that Zeller had done well to buy the tools from Roach. Such tools as the Philadelphia firms had offered at reasonable prices the bureau had purchased, so there was no reason for accusations of favoritism, Isherwood insisted. The tools were in perfect order, since Zeller had purchased them with the understanding that they would be in such a state, he had observed them working in New York, and when installed in Philadelphia, they operated without trouble.

So far as Tatham's complaint was concerned, he was no more than a "mouthpiece in the interest of others," with "no knowledge of machinery or tools," Isherwood maintained. The Philadelphia firms and their agent Rowbotham were obviously the source of the trouble, he insisted, because they had been thwarted in their attempts to sell the tools at exorbitant prices to the government. So confident had they been of Rowbotham's ability to control the selection of the tools that they had actually started to manufacture the equipment which they expected him to order. Unexpectedly meeting with failure, they determined to apply their influence as "immediate constituents" of Congressman Kelley in order to obtain redress.[4]

Under pressure from the many forces hostile to his department, Welles quickly established a board of three naval engineers to examine the tools purchased by Zeller. Isherwood heartily approved of the men selected for the board, observing that they were unbiased, trustworthy, and the best possible experts for such an investigation. The board chairman, Chief Engineer W. W. W. Wood, was not only the senior man in the corps but was unequalled as a machinery expert either in or out of the Navy, Isherwood declared.

Congressman Kelley had no intention of relying on any expert advice which came out of the steam engineering bureau. There had been "fabrication of false tables and elaborate calculations producing false results" in official reports coming from the bureau, he insisted, so a full-scale congressional investigation was in order.[5] Taking his cue, the *Army and Navy Journal,* on October 3, 1868, declared that Kelley had demonstrated "very plainly that there has been gross culpability somewhere," and the steam bureau was the place to find the culprit.[6]

Selecting his witnesses and directing their testimony, Kelley conducted the hearings on the tool purchases through the summer and autumn of 1868.[7] His picked witnesses dutifully testified that the tools were defective and generally obsolete. No one, it appeared, had taken the time to examine the equipment closely, and Zeller had been particularly negligent in failing to supervise the purchases adequately. Taking this opportunity to expand the area of his investigation, Kelley questioned the entire tool purchasing program of Isherwood's bureau since the end of the Civil War. Why so many purchases, he inquired, and why so often from Roach?

Roach, his rival tool manufacturers knew, enjoyed the lion's share of the business, selling over $900,000 of the $1,500,000 worth of tools bought by the Bureau of Steam Engineering over a two-year period. That Roach had obtained these orders because his prices had usually been lower than his competitors appeared to be an insufficient reason for Kelley. He was convinced that an officer in the bureau had doctored the figures to make sure that Roach would be the supplier. Kelley demanded a full disclosure of bureau finances and operations in order to show that someone, either "a crafty subordinate" or the "incompetent chief," was guilty of favoritism and corruption.[8]

At the same time that Kelley sought for evidence of illegality, the board of naval engineers produced a report, dated July 14, 1868, which stated that the steam bureau was free from error. Buttressing their conclusion with the expert testimony of twenty-five machine tool experts, the naval engineers told Welles that the equipment purchased from Roach was "first class" in quality, that Zeller's selection of tools had been "proper and judicious," and that the prices he had paid were quite reasonable.[9]

Kelley would have none of this. He had determined to find fault, and no contrary evidence could sway him. John Roach, observing Kelley's conduct, became pessimistic about the purpose of the congressional investigation and voiced his fears to his friend, William E. Chandler, attorney for Roach and Zeller. "Congress did not intend the committee to find fault with the tools because they were not purchased from Mr. Kellys constituents," Roach stated, "but this seems to be the main reason of Mr. Kellys investigation. . . ." Observing that the whole affair had been started by parties disappointed in obtaining contracts, Roach concluded that "Mr. Kelly . . . is not now acting as a disinterested judge and will not give a disinterested report. . . ."[10]

John Roach was all too correct in his fears. On January 5, 1869, the Congressman from Philadelphia presented the report of his investigating subcommittee, and the Navy Department braced itself for the worst.

Kelley began by stating that the quality of the tools purchased, the reputation of Roach as a tool builder, and the absence of competitive bidding in determining who would supply the steam bureau all reflected "but little credit upon the integrity of the management of the engineers at the Philadelphia Navy Yard," or on "the conduct of the engineer-in-chief of the navy in approving requisitions for these tools. . . ." There had been an "immense disproportion" in favor of Roach throughout all of the steam bureau's tool purchases since the end of the war, and Kelley demanded an explanation. The tools were not needed, he insisted, since little work had been going on in the Navy yards and even less was likely to occur.[11]

So far as Zeller and Isherwood were concerned in this matter, their behavior had been "irregular and indefensible," Kelley declared, "and the whole proceeding detrimental to the interests of the navy." The bureau Chief and his associates should be haled before a "competent court" which would be empowered to investigate the tool purchases for the entire Navy Department, Kelley urged. The "future welfare of the service" demanded this action, and there was no question in the Congressman's mind that these engineers should receive "official condemnation" for their "reckless disregard" of business procedures and Navy Department regulations.[12]

Kelley's "violent attack" on the Engineer Corps in Congress was "grossly false and calumnious," Isherwood indignantly wrote to Fox, January 7, 1869. "Not a word is true & most of it is ridiculous," he told the former Assistant Secretary, and he asked Fox's aid in enlisting support in the Senate to block proposed legislation with which hostile congressmen such as Kelley had threatened Isherwood's own office.[13]

The rantings of the Philadelphia Congressman were not unopposed in Congress. Realizing that Kelley's attacks on the steam bureau had ulterior motivation, the House Naval Affairs Committee determined to conduct its own investigation of the tool purchases, and on February 26, it submitted a report to Congress which exonerated Isherwood and his associates. Drawing on the conclusions of the board of naval engineers which had previously investigated the tool purchases, the House Naval Affairs Committee asserted that "the allegations . . . of the Hon. William D. Kelley . . . are entirely without foundation," and that Isherwood and Zeller had acted with propriety in the conduct of their business.[14]

Few congressmen had any illusions about Kelley's motives. He had forced his partisan investigation, declared fellow Pennsylvania Representative Samuel J. Randall, "for no other purposes than those of a political character, to bear upon the recent election." Failing in his attempt to satisfy the Philadelphia tool manufacturers, Kelley then had presented to Congress "a most unfair and unjust report . . . not in any manner approved by the House Naval Affairs committee."[15] Refusing to call witnesses such as Roach who might aid Isherwood and Zeller, and often restricting the testimony of those witnesses he did call, Kelley had attempted to push through a cynical and obviously political attack designed to satisfy his own constituents, embarrass the Navy Department, and provide further ammunition for both line officers and Radical Republicans who sought reasons for defaming the Navy Department and its Engineer in Chief.

The "Zeller Tool Scandal" played a significant part in drawing attention to the troubles besetting the Navy Department as it faced the consequences of the elections of 1868. Zeller had appeared to be lax in his duty, no one else had assumed responsibility in the tool purchases, and there was evidence to indicate that the management of the engineering department at the Philadelphia Navy Yard had been desultory; but these were not the real issues of the investigations.

As Kelley sought evidence to substantiate the claims of his disappointed constituents, he determined to broaden his investigations in order to bring the entire policy of the Bureau of Steam Engineering into question. Isherwood had undeniably made large and continual purchases during a period of retrenchment, and an explanation was in order.

The Engineer in Chief's insistence on enlarging government manufacturing and repair facilities in order to avoid dependence on private contractors was already a familiar issue in the Navy Department. By this policy, Isherwood received little support from private operators who welcomed the government business which frequently came their way when the Navy Department could not handle its own construction. In addition, Isherwood may have cut back drastically on his civilian work force, but he still was reluctant to see his corps lose the influence which it had exerted on the Navy during the Civil War. If he enlarged the government yards, he could reasonably demand workers in order to operate the added equipment, and this work would come under the cognizance of his steam bureau.

A number of the engines which Isherwood had designed and had contracted for before the conclusion of the war had been taken from the private builders in various stages of completion when the department cut back on its operations. Isherwood, however, still hoped to complete these engines and store them for future use. By obtaining large machine tools, he believed that the government would be able to finish the work on this machinery. Realizing that an incoming administration would doubtless be unsympathetic with his program, he may have attempted to accelerate work on the incompleted machinery in the hope of finishing construction before a new secretary of the Navy might cancel it altogether.

There was never any evidence that Isherwood would personally profit from these tool acquisitions or from the work they might do, or that he had made some *sub rosa* agreement with Roach in order to line anyone's pockets. Dedicated to his engineering work and to the operations of his bureau, he only sought to get as much money and equipment for naval engineering as possible, regardless of any ill will he might arouse.

The Kelley investigation quickly broadened into a general attack on Isherwood and the department because there were so many groups which could benefit from the discrediting of the Engineer in Chief. Line officers could use it to attack Isherwood's supposedly corrupt regime; private contractors could employ it in their attempts to eliminate Roach as a competitor and to seize government work for themselves; Radical Republicans could take advantage of it to condemn Welles and his associates who had refused to support the impeachment of President Johnson, and who demonstrated

Democratic sympathies which the "bloody-flag" Radicals would not tolerate.[16]

In the midst of these troubles, Isherwood tactlessly renewed his dispute with John Ericsson. In his annual report for 1868 to Welles, the Engineer in Chief maintained that he had invented the system "now universally employed in ventilating the monitors." The original *Monitor* had been "utterly deficient" in ventilation, he explained, and her galley had to be placed on deck because the interior of the vessel had been "uninhabitable." Ericsson had refused to alter the arrangements of his vessel, characteristically ignoring the practical problems of operating his own warships. After her battle with the *Virginia,* the *Monitor* had been taken to the Washington Navy Yard where Isherwood had improvised a ventilating apparatus which proved to be "a complete success," according to the Engineer in Chief.[17]

When the Secretary of the Navy's report was published in December, 1868, Ericsson and his associates became furious. The inventor's close friend, Isaac Newton, branded Isherwood's claim "a fitting climax to his history of the screw Propeller, his attempt to steal the plans of the original 'monitor' and from them to construct a fleet of turret ships," not to mention Isherwood's subsequent attempts to harass Ericsson. "He is clearly entitled to a sound drubbing from your hands," the engineer insisted.[18]

Newton, hired by the line officers to produce newspaper articles and pamphlets assailing Isherwood and his *Wampanoag,* apparently had personal reason to attack the Engineer in Chief. According to Isherwood, Newton had been interested in the design and construction of the light-draft monitors back in 1863. He had presented working drawings of a small machine he had patented to the builders who "innocently worked from them," assuming that they were department designs. After the vessels had been completed, Newton presented a bill for $1,000 to each contractor for his patent fee, expecting to receive a total of $20,000 by this clever stratagem. The contractors appealed to the Navy Department, which directed that the fee should not be paid. Newton assumed that Isherwood had been responsible for thwarting him, and ever since he had held the Engineer in Chief in his debt for the $20,000.[19]

Furious at Isherwood, Newton now urged Ericsson to attack the bureau Chief in print. "I hope that in addition to the castigation you are about to give Isherwood in a communication to the Secretary," Newton wrote the inventor, January 6, 1869, "you will write an editorial article on this bad man, either for the 'Times' or 'Army and Navy Journal'...." Newton suggested that Ericsson contribute an article to *The New York Times,* since they had an agent who was now associated with the newspaper, "and anything he presents, *goes in—*...."[20]

Ericsson first wrote to Welles, January 12, 1869, insisting that his ventilating system for the original *Monitor* had worked and demanding that the Engineer in Chief retract his statements.[21] The Secretary turned the inventor's protest over to Isherwood, who replied to Welles in a detailed letter several days later. Ericsson, he said, had never figured out a way to ventilate his monitors effectively. After the battle with the *Virginia,* the *Monitor's* crew had to come up on deck and remain there to escape being suffocated inside the hull. Although the *Virginia* still threatened the Federal vessels at Hampton Roads, the *Monitor* had to be towed two hundred miles up to Washington to receive a proper ventilating system. When Ericsson and Stimers failed at that time to correct the deficiencies, Isherwood undertook the task "with extreme reluctance," but "immediately perfected the system . . . now universally employed in ventilating the Monitors."[22]

Ericsson, explained the Engineer in Chief tartly, merely had a "defective memory" which "probably confuses dates and facts," since the inventor had certainly failed to invent a working ventilating system for his vessels. Isherwood had been forced to deal too often with inventors to excuse their eccentricities. "After a lapse of a number of years," he told Welles, in a letter dated January 18, 1869, "it is difficult for an inventor to believe there is anything about his original invention, however afterwards perfected by others, which he did not invent himself...."[23]

In the meantime, the press had already joined the dispute. On December 26, 1868 the *Army and Navy Journal* declared that it could not print the words which would properly describe Isherwood's report and pronounced it to be "the last literary monument, probably, of its prolific author's career."[24] *The New York Times* was especially disturbed, because English journals, which for so long a time had ridiculed Isherwood with unsparing severity, now congratulated him for depreciating the monitor system which was so criticized in Europe. Isherwood, *The New York Times* asserted, January 27, 1869, had committed an unforgivable breach of honor by his "ungrateful and unpatriotic task of decrying the American monitor system to which we owe so much in the past and whose world-wide prestige is our security for time to come." By daring to question the work of Ericsson, Isherwood had committed an "official discourtesy" which was especially reprehensible now that "the navies of all maritime powers are controlled by [monitor] principles, . . if not already reconstructed on its plans."

To such protests Isherwood was as indifferent as he had been to the numerous pamphlets which assailed his own abilities. The "lucubrations" of the *Army and Navy Journal* and of the numerous pamphleteers were "mere twaddle," he informed Fox, January 5, 1869. Invariably their facts were false and their inferences "simply ludicrous." The criticism he had received from English journals he ignored, recognizing that many of the articles quoted in American papers from *Engineering* and the *Engineer* had actually been written by such men as Ericsson, and then sent to London in order to be printed and to appear as impartial authority.[25]

Refusing to be bothered by most of his critics, and contemptuously dismissing those whom he chose to answer, Isherwood could not dispel the general impression which they were able to create about his abilities and his conduct as bureau Chief. At a time when line officers were out for his blood and politicians recognized his value as a scapegoat, he needed friends more than ever within his own profession if he were to defend himself. By his poorly timed and disparaging comments on Ericsson and the monitors, Isherwood only alienated one more group and added them to the swelling number of those who now actively worked to rid the Navy of its Engineer in Chief.

In Congress, William Kelley had allied himself with civilian engineers and line officers who saw in the office of the chief of the steam engineering bureau the greatest burden on the Navy. On the same day that he presented his report to the House on the Zeller tool purchases, Kelley brought out of committee a bill which would make provision for a civilian engineer to be appointed chief of the steam bureau.

This was an issue of long standing in the Navy Department.

When Isherwood had first been appointed in 1861, he had joined with Senator Grimes to defeat an attempt to open up the office of engineer in chief to civilians, the third time that such an effort had been made. Grimes was then able to push through legislation in August, 1861, which restricted the position to naval chief engineers, but in the following year the House Committee on Naval Affairs again attempted to bring in a civilian when the Bureau of Steam Engineering was created. Amending the Senate bill on naval reorganization to read that the office of chief of the steam bureau would be open to any "skillful engineer," the House failed to achieve this goal when the Senate Naval Affairs Committee refused to concur and persuaded the Senate to vote against the proposal.

Continuing pressure by civilian engineers to gain entrance to the steam bureau received impetus when Representative Robert T. Van Horn, of Missouri, at the instigation of Admiral Porter, introduced a resolution in the House in February, 1868, calling for legislation which would allow the President to appoint a civilian head of the steam bureau.

Considering this as the "last effort to dislodge me," Isherwood gloomily prophesied to Fox, on February 18, 1868, that if the bill passed, a civilian would immediately be appointed in his place. "I can only say," remarked the Engineer in Chief with dour satisfaction, "that my illustrious outside successor, whoever he may be, will find it difficult to beat the Wampanoag."[26] Isherwood had been too pessimistic; once again the effort to bring in a civilian failed to obtain immediate congressional approval, as Van Horn's bill remained dormant in the House Naval Affairs Committee.

However, after the elections of 1868, Congress was no longer so protective. The notoriety of the Bureau of Steam Engineering's operations, springing from the tool investigations and from the efforts of David Porter, strengthened the hand of civilian engineers and impelled Congress to action. When Kelley presented the results of his tool investigations, he also brought Van Horn's bill out of committee and, in a stirring address to the House of Representatives, recommended its passage.

"It is not believed by the Naval Committee that our Navy exhibits the highest character of engineers," Kelley began, adding that other navies utilized both steam and sail, while the American Navy, under the influence of its engineers, wastefully used steam alone.

Despite this dedication to steam, American warships were slower than those of other navies, "even of those of the little South American States which have navies," the Congressman scornfully continued. This reprehensible failure of American naval vessels was caused by the fact that a naval engineer, rather than a civilian, had been in control of the Bureau of Steam Engineering, thus wasting hundreds of millions of dollars by utilizing impractical and inefficient machinery.

We have the slowest Navy in the world; that in the strife between the engineer corps and the line the engineer corps, endeavoring to exalt itself above the line, have made our ships depend entirely upon steam, fuel, and machinery, so that the engineer shall be the important man on the ship, and subordinate to him the naval officer to whom the command belongs. . . .

The blame for this deplorable situation, Kelley asserted, should fall on Isherwood. This "cherished head of the engineer corps," he assured his fellow congressmen, had demonstrated little ability, since his *Wampanoag* could not carry sufficient fuel or provisions to be an effective cruiser, and was notorious for its cramped officers' quarters. Unfortunately, Kelley said, Isherwood could not be easily disentangled from his corps. There was an annoying "esprit du corps," he noted, which had resulted in the engineers' support of their Chief over the "stupid or corrupt transactions" involved in the Philadelphia tool purchases. He believed, therefore, that the whitewashing of Isherwood and Zeller by the board of engineering experts was either "complicity in crime" or a misplaced "esprit du corps." In either case, Kelley decided, it could not be condoned, and a civilian must replace Isherwood in order to break up this engineering ring.[27]

Viewing the furor over the steam bureau and Isherwood, Representative Elihu B. Washburne expressed an increasing congressional sentiment by testily declaring, "this Bureau . . . has been the source of more trouble, complaint, and expense to the Government than all the other bureaus. . . ." The extent of this sentiment became disturbingly apparent to Isherwood when the House voted on Van Horn's bill for a civilian engineer and passed it ninety-seven to twenty-three.[28]

David Porter was jubilant. Writing to a fellow line officer that the bill had passed the House by a "six to one" majority, thus "consigning Mr. Isherwood to the tomb of the Capulets," he predicted that the bill would pass the Senate almost unanimously.[29] *The New York Times* joined in the celebration by publishing a long editorial, January 12, 1869, which severely reviewed Isherwood's career, applauded Kelley's speech, and looked forward to further disclosures with which the Philadelphia Congressman had tantalized the press. He had promised a "vast mass more damning," and the enemies of Isherwood chafed with impatience for its appearance.

Despite Kelley's support, Van Horn's bill, as the others before it, failed to pass successfully through the Senate. It remained in the Naval Affairs Committee for several weeks until Senator Grimes received permission from the Senate to have his committee discharged from further consideration of the bill. Isherwood, however, was by no means free from congressional attack. In the previous months, David Porter had worked assiduously in persuading Senator Grimes of Isherwood's faults and a need for naval reorganization; and in March, 1869, the Admiral's efforts bore fruit. Grimes introduced into the Senate a bill which called for a Board of Naval Survey, to be composed of three line officers, all of rear-admiral rank or above. This triumvirate, staff officers believed, would virtually abolish the bureau system and would establish a "military protectorate" over the Navy Department if it came into power. As this group by law was to hold office for only three years, there would inevitably be a triennial power struggle in which the civilian secretary would be helpless, because of his subordination to the line-officer control of the department.

"It is well understood," Welles noted in his diary, that Porter was to be president of this Board of Admiralty. This would place him in the Navy Department "as superior or superintendent," and allow full line-officer control, the Secretary believed. It was a scheme, he later observed, which had long since been planned. Porter had not only insinuated himself with Grimes, largely through Walker, Grimes's nephew, but the Admiral had also played up to the President-elect, Ulysses Grant. During the 1868 campaign, Porter had testified "that Grant was a total abstinence man," Welles noted with disgust.[30] Moreover, when Grant visited Annapolis a few days after his nomination, Porter had provided two days of great demonstration, a notable change from the prenomination days when Grant had received no special attention from the Admiral.[31]

Porter, Welles had long since decided, was a schemer and intriguer who obeyed no other principle than making sure that he was on the winning side. Before Andrew Johnson's impeachment, Porter had been openly a friend and supporter of the President; but with the impeachment he suddenly turned strongly against the entire Johnson administration, "scandalizing and abusing" the President with vigor and apparent relish. At a time when Johnson's conviction had seemed certain, it was rumored that Porter had made an arrangement with Senator Ben Wade whereby Porter would become Secretary of the Navy if Wade obtained the Presidency. Then, if Grant succeeded Wade, Porter would continue in control of the Navy Department, consolidating the power of line officers and, especially, that of Admiral David Dixon Porter.

As the inauguration of Grant approached, the fate of Welles's naval administration became increasingly apparent. With support from Kelley in the House and from Grimes in the Senate, and backed by the power of the line officers, Porter drew nearer to the seat of power. "I pity the next Secty. of the Navy," the ailing Grimes wrote to Fox in discouragement, on January 28, 1869. "The whole institution is in a demoralized condition. No money will be appropriated, nothing will be done & his sole employment will be the settlement of controversies...."[32] Disgusted with the Navy, and perhaps with himself for aiding in the line officers' grasp for power, Grimes decided to resign his place on the Senate Naval Affairs Committee, and only with reluctance did he retain his seat in the Senate, for he wished strongly to be gone when the new administration and Congress assembled in March.

Within the corps of naval engineers, Isherwood failed to command the loyalty of all of his associates. The likelihood of his removal if Porter took over the Navy Department proved to be an irresistible lure to at least one engineer who, for years, had jealously eyed Isherwood's position.

Chief Engineer James W. King not only believed that Isherwood should be removed as bureau Chief, but also was convinced that he was the logical choice as successor. He had disagreed with Isherwood's steam theories during the early 1860's, and in 1862 he had done everything in his power to defeat the confirmation of Isherwood as bureau Chief, in order to obtain the position for another chief engineer with greater seniority.

A slim, subtle man with lazy eyes and a straggly mustache

flanked by luxuriant mutton-chop whiskers, King strongly contrasted in appearance with Isherwood. In personality they were equally dissimilar. Isherwood was notorious for his direct, gruff manner, and his intolerance of those he believed to be in error was as intense as his personal and professional honesty. King, on the other hand, was "smooth and never uses harsh language," according to Secretary Welles. Inferior to Isherwood in mental capacity, King attempted to make up for this deficiency by his manipulation of friends and enemies alike. "Those who differ with him," Welles noted, "charge him with plausibility and insincerity. . . ." An ambitious man and a schemer, James King fitted in well with Admiral Porter.[33]

King, who would later be accused of aiding Kelley in the tool-purchase investigation in order to obtain the position of engineer in chief, now began actively to solicit aid.[34] On February 7, 1869, he addressed a letter to Gustavus Fox, who, although no longer assistant secretary of the Navy, had retained his contacts and his interest in the affairs of the department. "There can be no doubt," King began, "but that Mr. Isherwood will be relieved from the Bureau next month. He has been there too long, so long that his personal unpopularity has operated seriously to the detriment of the whole Corps." Not only did the engineers desire to be rid of their Chief, King maintained, but line officers, "knowing citizens," and several senators were determined to see this change.

As chief of the steam bureau was the only further promotion open to him, King explained, he naturally coveted the position. Because numerous friends, including three senators, had urged him to obtain this office, he planned to visit Washington by March 4, "to apply for the position and to remain there until the question be decided."

Fox, acknowledged King, had the greatest knowledge of the Navy and its officers, in addition to having a thorough comprehension of the "political workings of Washington society." Since Fox also had important contacts, both in the Navy and in Congress, King explained that he had logically written to him to enlist his aid and to request a letter of recommendation. Knowing that Fox wielded great influence in Washington, King frankly admitted, "I do not hesitate to ask it."[35] The inauguration of President-elect Grant was still almost a month away, but King would take no chances. With Porter and Fox both behind him, he reasoned that he should have no difficulty in displacing Isherwood.

Isherwood may have been no politician, but he was no fool. The inexorable march of events in Washington had become clear even to one so deeply immersed in his work and usually indifferent to maneuverings for power. Becoming concerned for his position as Chief of the Bureau of Steam Engineering, he began to probe for support by turning to the Secretary of the Navy who had stood behind him for so many years. Requesting a statement of department policy which would defend his actions on the tool-purchase question, Isherwood suddenly found the Secretary evasive and unwilling to commit himself in defense of the Engineer in Chief. His policy, Welles told Isherwood, in a letter dated February 17, 1869, had always been to encourage additions and improvements to the Navy yards, and for many years he had prominently featured in his reports to Congress the need for money to improve inadequate facilities. However, he quickly added, the responsibility for providing the means to repair naval machinery was Isherwood's, and however he may have chosen to do this was entirely his own affair. Disclaiming any knowledge of the details of tool purchases for Navy yards, Welles implicitly refused to make a formal statement defending the Engineer in Chief from the mounting attacks in Congress which had grown out of the Zeller tool purchases.[36]

Realizing now that he had little effective backing within the department, Isherwood determined at least to prevent the incoming administration from undoing work which he had in progress. In an attempt to insure that the numerous engines of his own design which were in process of construction would not be lost to the department, he sent a circular letter to contractors stating that if they had finished work up to a certain point, he would accept the engines at that stage and would deduct only $40,000 from the total payments the government had contracted to make to the builders. Unwilling to lose these funds, contractors refused to accept Isherwood's terms and continued their desultory work on the engines.

And then it was too late to do anything more. On March 4, 1869, Ulysses S. Grant took office as president of the United States. On the afternoon before, Gideon Welles had packed his belongings and departed from the Navy Department while Grant prepared to appoint his own secretary of the Navy.

Wary of appointing a career naval officer to fill a cabinet position traditionally held by civilians, Grant cast about for someone to utilize as a figurehead, while enabling Admiral David Porter to exercise effective control of the Navy. In Adolph Borie, Grant found his man. A genial Philadelphia merchant, Borie had no particular interest in the Navy nor was he eager to take on the job vacated by Gideon Welles.

This proved to be no hindrance, however, as Grant telegraphed Porter on March 9 to come from Annapolis to take charge of the Navy Department. With his mind bulging with general orders which he could scarcely wait to impose upon the Navy, Porter quickly came to Washington on a special train. Met at the railroad station by Borie, who had been waiting anxiously for his arrival, the Admiral and the new Secretary went "arm in arm" to the Navy Department where Porter announced to Assistant Secretary William Faxon that he had arrived to "run the Department." When Faxon, out of habit, began to inform Secretary Borie of the office routine, Porter peremptorily interrupted and told him that henceforth all information and communications to the Secretary would be channeled through Porter. "In all this," Welles noted in his diary, "poor Borie was a passive tool," and it was obvious to all that he had come to the department only to be "a mere clerk" to the ambitious Admiral.[37]

With Porter in control of the Navy Department, Isherwood turned to Fox as the only person who might offer him aid. Noting, in his letter to Fox of March 9, 1869, that Porter was in Washington "to run the machine" for Borie, Isherwood still hoped that the Admiral would prove to be only a temporary aggravation, since it was rumored that Commodore Daniel Ammen was on his way from the East Indies to become "permanent attache" in the department. However, Isherwood reported to Fox, "the enemy has already begun the attack on the new Secretary for my removal." Borie, surprisingly, had given Isherwood this news himself, apparently reluctant to obey the dictates of Porter as they applied to the Engineer in Chief. "His reluctance," observed Isherwood drily, "will no doubt be overcome."

However, there was still one thread of hope, Isherwood continued. "Everybody knows that Porter's fame & fortune are the work of your hands, and it is just possible he may be disposed to gratify you in some trifle in return. He ought to do anything you ask," Isherwood added, in a pathetic attempt to build up his own meager confidence. "If you consistently can, will you write him in this matter, and engage him, if possible, on my side?" he inquired. After all, Isherwood pointed out, his normal term in office would not expire until one year from the following July,[38] and if he were dismissed now, while the other bureau chiefs remained, it would be embarrassing and would imply that there had been a valid cause for his dismissal. He had no desire to stay in the department beyond the normal expiration of his tenure as bureau Chief, Isherwood assured Fox. Nevertheless, he believed that he was "fairly entitled" to finish out his term of office honorably. To make this possible, he added, "it will be necessary to be prompt."[39]

Whether Fox was prompt or not would make no difference in respect to Porter, whose implacable opposition the Engineer in Chief curiously refused to admit. Not unfriendly towards Isherwood, Fox nevertheless realized that it was too late to do anything to help him. Just as there had been no necessity to give King a helping hand to the office he prized, there was now no reason to make a vain attempt to halt the purge which was about to sweep through the Navy Department. David Porter was in control, and it was useless to attempt to impede him.

Porter took over the Navy Department with his customary flair. He immediately disbanded the board of engineers which had cleared Isherwood and Zeller on the tool purchases, and substituted in their place another Board of Inquiry with orders to examine the entire operations of the Bureau of Steam Engineering. He demanded a complete accounting for every engine built since Isherwood had come into office in 1861, and he called special attention to those not constructed under proper congressional authorization. He ordered that the board scrutinize the purchase of all machine tools and analyze critically all steam bureau expenditures. The board, Porter stated, was also to investigate all people employed for whatever purpose by the steam bureau in addition to all those in any way connected with the bureau's contracts.

Having dealt with the tool-purchase affair, Porter next turned to the staff officers. To provide an unmistakable example and warning for those who would oppose the line, he ordered that a naval constructor, currently occupying a house in the Boston Navy Yard, immediately vacate the dwelling, since it had been built originally for a line officer and "its present occupancy causes dissatisfaction." Within six days, the naval constructor had been removed, and in his place appeared a line officer, in accordance with the Admiral's decision that only these officers should be allowed to enjoy such a privilege.

It was not sufficient for Porter to be the power behind the throne. Within a few days after coming to the department, he decided that it was unnecessary and impractical to mask his own dictates with Borie's apparent authority. On March 13, he therefore issued over Borie's signature an order to all commandants of Navy yards and stations, informing them that from that date they would "recognize all orders coming from Vice Admiral Porter, as orders from the Secretary of the Navy."[40] His way was now clear to begin the naval reorganization he had so diligently planned and so impatiently awaited.

In the meantime, the Admiral had wasted little time with Isherwood. On March 10, President Grant nominated Chief Engineer King as chief of the Bureau of Steam Engineering in place of Isherwood, "whom I desire removed," the President explained. On the same day, King's nomination went from the floor of the Senate to the Committee on Naval Affairs, where it received immediate approval. Returned promptly to the Senate, King's nomination received formal confirmation, thereby making his appointment official and marking the conclusion of the tenure of Isherwood as engineer in chief.[41]

"Porter has begun his career by an onslaught on Isherwood," Welles observed. Reflecting upon the career of the controversial Engineer in Chief, the former Secretary of the Navy admitted, in his diary, that "Isherwood has his peculiarities," but decided that he had offset these by being "mentally superior to any one of the chief engineers with whom I have come in contact." Although he had not displayed great talents for business, Welles noted, Isherwood had been devoted to his profession and had produced engines which had rendered good service and had proved to be superior to any others, despite the denunciations by Isherwood's rivals and critics.[42]

Isherwood's error, Welles decided, had been to lead his corps, and staff officers in general, in their "clamor for rank." Not an accomplished politician, Isherwood lacked that *"suaviter in modo"* which

Welles recognized as such a valuable asset in dealing with controversial subjects. Instead, with his ready and prolific pen, Isherwood had spoken his mind, "roughly and offensively at times," and his indifference to his opponents' accusations had betrayed him.[43]

On March 16, Isherwood received his orders to leave the Navy Department and turn over to his successor all the records of his bureau. Assembling his personal papers and carefully packing the voluminous experimental data which he had collected during the previous years, he prepared to leave for his home in New York city. No longer Engineer in Chief of the United States Navy, no longer head of his own bureau, he now was merely one of many chief engineers who would patiently serve out their remaining years in a Navy which had turned its back on progress and had lost its status as a world power.

Not yet fifty years old, Isherwood faced a future life far different from the one he was about to leave. For eight years he had enjoyed authority and responsibility as the engineering expert of the American Navy. Now he was to be discarded and, his opponents devoutly hoped, forgotten. Nevertheless, he knew that he had left a mark of engineering achievement, both in theory and in practice, which could not be erased. Above all, he had unremittingly dedicated his energies and abilities to his country in time of crisis. The Union Navy had called on Benjamin Isherwood for steam power, and the engineer had supplied it.

## XII. Always the Engineer

When he was relieved of the duties of Engineer in Chief, Benjamin Isherwood had not yet reached his forty-seventh birthday. He was to remain on active duty until 1884, when he retired at age sixty-two; he then would pursue a busy and productive career as writer and experimenter well into the twentieth century. The second forty-six years of his life perhaps do not match in historical interest the earlier career which had culminated in his services as Engineer in Chief; yet for Isherwood, these remaining years may have been as rewarding as the others and, after his retirement from the Navy, may have been the most enjoyable of his professional life.

During the first few months of the Grant administration, Admiral Porter busied himself with the task of turning the clock back to the age of sail. Isherwood and the steam Navy which had been built under his direction were not ignored. Shortly after removing him from office, Secretary Borie ordered Isherwood to prepare reports on all experimental work done in steam engineering during his tenure as Engineer in Chief. When it came to evaluating the design and performance of Isherwood's vessels and machinery, however, the new naval administration under the guidance of Porter would rely largely on the judgment of line officers and discount the opinion of naval engineers—especially those who had worked with Isherwood.

Under the direction of Rear Admiral Louis M. Goldsborough, the board of investigation established by Porter, in March, spent several months visiting one Navy yard after another to make an examination of all "steam machinery afloat." Since Isaac Newton had been hired as an expert civilian engineering advisor, the nature of the majority report submitted by the board in late September was predictable. Overriding the feeble objections of the two naval engineers on the board, the majority of the members produced a sweeping condemnation of all vessels built under the direction of Isherwood. Their recommendations for "improving" these vessels, which were eagerly accepted by the Navy Department, went a long way toward achieving the goals of line officers and their leader, Admiral Porter.[1] Along with her sister ships, the *Wampanoag* subsequently suffered not only the indignity of a name change but also an overhaul which removed half her boilers, loaded her down with masts and yards (to achieve full sail power), and turned her into a useless weapon of war. Isherwood's masterpiece, renamed the *Florida*, spent the next fifteen years rotting at the wharf until she was finally sold, in 1885, for $41,508—a pitifully small amount for a vessel that had cost over $1,500,000 to complete and had been a landmark in the development of American naval architecture and engineering.

Isherwood, unlike his cruiser, was not to moulder in the service until retirement. Engineer in Chief or not, he still considered himself to be an expert naval engineer, and if he was to be "exiled" from Washington, then it was proper that he should pursue his professional career elsewhere as constructively as possible. Ordered in July, 1869, to a humdrum assignment at the Mare Island Navy Yard in California, Isherwood quickly requested a six-months' leave of absence so that he might make a "professional visit" to Europe. Although Borie had been replaced as Secretary of the Navy by George Robeson, in June, the department still appeared to be under Porter's influence. Isherwood's plan to improve his knowledge of foreign engineering and naval architecture, even if at his own expense, fell on unsympathetic ears, and by September he had reluctantly arranged his affairs at home and had departed for duty in California.

Once he had reached Mare Island, Isherwood made the best of his situation. While conducting extensive tests with the various coals of the Pacific Coast region, he also measured the performance of several screw propellers of various designs in order to determine their comparative efficiency of operation. The data which he compiled and later published from these propeller tests was so thorough and exact that it would, subsequently, be hailed by English and American engineers as basic for the work of scientists who over the next generation would develop original propositions in propeller design and propulsion theory. Characteristically preferring to contribute factual data to support the original work of others rather than strive to be an inventor or creative theorist himself, Isherwood was not attempting to avoid controversy. Indeed, now that he had sufficient time, he wrote to Lenthall, he intended to write books on steam engines which he fully expected to "keep him in hot water for the remainder of his life."[2]

Benjamin Isherwood was not only an engineer, he was a naval officer, and despite a general department order in March, 1869, which had further reduced his relative rank to commander, was senior to all line officers but the commandant at Mare Island.

The line-staff controversy was still raging when Isherwood arrived at the California Navy yard; this was undoubtedly a factor in producing the situation at the end of 1869 when the Mare Island commandant, Rear Admiral Thomas T. Craven, declared himself incapable of giving Isherwood the usual quarterly fitness evaluation, being "too much prejudiced against him to make a fair report." Although Craven was relieved in 1870 as commandant by Commodore John S. Goldsborough, Isherwood evidently remained under a cloud, since Goldsborough twice placed the engineer below all other staff officers in the Navy yard in the category of "morals," and despite Isherwood's research and experimentation programs, Goldsborough gave him mediocre grades for his professional aptitude, zeal, capacity, and display of general information.

The relations between Isherwood and the line officers came to a head late in 1871 when a new commandant, Commodore Enoch G. Parrott, complained to Secretary Robeson that Isherwood outranked the executive officer at the yard and had thus refused to observe many of the forms of naval etiquette—especially in neglecting to inform the executive officer when he was departing from and returning to the yard. As this created a very embarrassing situation for the line officers, Parrott asked Robeson to transfer the engineer elsewhere. Within a few weeks Isherwood had been detached from duty at Mare Island and was on his way home to New York city.[3]

Isherwood spent the next year working in his office at home with his companion Theodore Zeller and serving as a senior member of a variety of engineering boards of investigation. However, in January, 1873, the Navy Department sent him to Key West, Florida, as a consulting engineer in the construction of a foundry. Unhappy with this assignment and wishing to return to the more congenial duty in New York on the boards of investigation, Isherwood wrote to Robeson in April, requesting reassignment on the basis that his professional ability was being wasted in Florida. Although he failed to return to New York in this manner, he was back home by July for another reason—a recurrence of the "chronic Asiatic dysentery" which had plagued him in the 1850's.

No longer a young man, Isherwood remained sick for over a year, despite hospital and home care, before he could shake off this debilitating illness; and then in September, 1875, when he had barely recovered, the Navy Department ordered him to Yokohama, Japan, where he was to serve as supervisor of engineering repairs for the Asiatic Squadron. Appalled at this prospect, Isherwood begged for an extension of time because his mother, now past seventy-five years of age, was lying "at the point of death with the contest for life still dubious." As he was her only child, Isherwood informed Robeson, he had to arrange her property settlement. Furthermore, he added, conveniently ignoring the fact that his wife lived with his mother, his several children would be left without anyone to look over them, since his mother had always been "their only protector in my absence."[4]

Isherwood's barrage of letters and telegrams finally wore down the resistance of the Navy Department which granted him a delay so that he could care for his dying mother and his children. Two weeks later, Isherwood asked for a one-year leave of absence so that he might travel to Europe for a tour of professional study of marine steam machinery; incredibly, the Navy Department obligingly canceled his orders for the Far East and granted his request.

Isherwood and Theodore Zeller spent the next eighteen months traveling throughout Europe. That their tour was not entirely for professional purposes is apparent from the letters to Lenthall in which Isherwood grandiloquently described the cultural wonders of Greece and Rome, but he was sufficiently practical to avoid paying his own way any more than was necessary. By early February, 1876, both he and Zeller had managed to obtain orders for "special duty" in Europe, for which purpose they could both draw their regular pay and collect traveling expenses. For the next fourteen months the two visited every Navy yard in France and Great Britain, collecting data on hulls, armor, machinery, and the performance of every warship they could find.

Returning to New York city in April, 1877, Isherwood resumed his duties on the experimental boards while he prepared his report on the European tour which he presumably submitted to the Navy Department late in that year. For the remainder of his naval career he was to continue serving, usually as senior officer, on various special naval boards which investigated the structure and performance of steam machinery and steam vessels. Such was the thoroughness of Isherwood's work that his office at home rivaled the files of the Navy Department as a repository of data; more than once, especially after the occasional fires which ravaged the Navy Department files, Isherwood became the sole source of engineering information for the department.

No longer responsible for the design and performance of naval machinery, Isherwood still was a controversial figure. Aside from continuing to be the whipping boy for the *Army and Navy Journal,* which blamed Isherwood throughout the 1870's for any problems in American naval engineering or warship design, he managed to involve himself in political disputes both in and outside of the Navy Department. Together with his friend John Lenthall, Isherwood became a formidable expert on the deficiencies of naval design and administration throughout the years of Republican control from the presidential administration of Grant to that of Arthur. In 1872, Isherwood was brought before a Republican-controlled House investigating committee which sharply questioned his public criticism of the Navy Department's contract procedures for vessel repairs. By 1878, however, the Democrats were in control of the House, and Isherwood and Lenthall accused both the Bureau of Steam Engineering and the Bureau of Construction and Repair of a "flagrant exhibition of gross ignorance and culpable carelessness." Isherwood paid particular attention to the Chief Constructor Isaiah Hanscom, whom he accused of making "preposterous, ridiculous, and ignorant statements. It is not safe," Isherwood concluded in his characteristic manner, "for any man to limit the knowledge of others to the horizon of his own ignorance."[5] Understandably, Robeson insisted that Isherwood could not be trusted as an expert; "for the last ten years," the Secretary insisted, Isherwood had been "the Democratic expert in these attacks on the Navy Department."[6] Yet Isherwood's criticisms were probably not so much out of revenge or partisan political motives; he was a perfectionist, and the Navy of the 1870's fell far short of perfection, regardless of the cause. Contract procedures *were* suspect; construction policy *was* unenlightened and wasteful; and the state of American naval engineering as well as warship design had unquestionably slipped far behind the work being done in Europe. Explanations and extenuating circumstances were plentiful, but Isherwood was concerned only with the state of his art, which was undeniably ripe for criticism.

As a measure of their professional stature, especially in view of their opposition to Republican naval administration, both Isherwood and Lenthall were selected for membership on the First Naval Advisory Board, established by Secretary of the Navy William H. Hunt, in July, 1881, to provide an organized and feasible plan for building a modern American Navy. The rebirth of the Navy, its emergence from the "Dark Ages" of reactionary commitment to outmoded strategy, tactics, and weapons of war, had its first real stimulus at this time; even here, however, the role Isherwood was to play would be one of dissent.

Split by serious differences of opinion, the board submitted both a majority and a minority report to Secretary Hunt, in November, 1881. That the Navy should embark on a program of expansion and modernization was accepted by all; beyond this point, however, there was disagreement over the size, speed, armament, and armor of vessels; the size and power of engines; the amount of sail power; materials of construction; and above all, the basic purpose of the warships to be recommended for construction. Isherwood and Lenthall, in the minority, reiterated their position of 1862, arguing for the ultimate necessity of heavily armed and armored, seagoing ironclads rather than for the monitors, rams, and small torpedo boats which the majority still considered appropriate for a passive policy of coast defense. Since the limited availability of funds made his ironclad program impractical, Isherwood recommended building a *Wampanoag* type of commerce destroyer as a stopgap measure; even here, however, he was not to be satisfied, since the majority preferred unarmored cruisers of a more modest dimension, especially in respect to engine power.[7]

At one point, both Isherwood and Lenthall adopted a position which, in light of future developments, has been considered unrealistically conservative. On the issue of material for hull construction they opposed the use of steel and insisted that iron was the preferable material. Basing their position on the existing limits of metallurgical technology, they argued that "steel" as it was then produced in the United States was at best a high-quality iron, relatively untried as a shipbuilding material, unpromising in its characteristics, and much more costly than iron. Fearing that the Navy might commit itself to a prohibitive expenditure by underwriting the development of the infant American steel industry, Isherwood and Lenthall took a characteristically practical position which they considered both realistic and constructive, in view of the existing state of warship production.

Disregarding the warnings of Isherwood and Lenthall, the Navy Department called a second Advisory Board in 1882, on which they were not asked to serve. From this board came the recommendations which eventually produced the first ships of the "new Navy"—three unarmored cruisers and a despatch boat, familiarly referred to as the "ABCD" ships. These vessels had hulls of steel, considerable sail area, and speeds and cruising ranges which fell distressingly short of Isherwood's ideal for an unarmored commerce destroyer. Built under a Republican administration, and by a contractor, John Roach, who had close ties to Secretary of the Navy William E. Chandler, these vessels were vulnerable to attack both for professional and political reasons. Once more, both as Democratic expert and engineering critic, Isherwood served the purposes of those who sought to condemn these four vessels as deplorable products of a hopelessly incompetent if not corrupt Republican administration.[8]

Granted his political involvement, Isherwood's solicitation, in June, 1883, of the temporarily vacant position of engineer in chief displayed either a remarkable degree of temerity or an equal amount of shortsightedness. Writing directly to President Arthur, June 6, 1883, for the appointment, Isherwood assured him that the office of engineer in chief held no political importance and that, as senior naval engineer in the service, he had at least a moral, if not a legal claim to the position. Futhermore, his professional accomplishments, qualifications, and world-wide reputation should dispel any further doubts about his being the right man for the post. "As an engineer in the highest sense of the word," Isherwood flatly stated, "I believe it is conceded that I have no superior, and in my own corps most certainly no equal."[9]

Isherwood failed to regain his old office—by which he could have retired the following year at higher rank and pay, as Republican critics were quick to point out—but he was not alone in believing that he should once more hold it. Surprisingly, a staunch supporter by this time, whose opinion of Isherwood would continue to rise throughout the 1880's, was Admiral of the Fleet David Dixon Porter. The burning issues of the late 1860's having subsided, the two men became reconciled; Isherwood was now the engineering advisor for Porter's torpedo-boat schemes; and the Admiral, in return, had come to believe, by 1881, that the incumbent steam bureau chief should be removed and be replaced by Isherwood, "the only man who can at present fill the place with benefit to the navy."[10]

Isherwood concluded his forty years and eight months of service in the Navy on October 6, 1884, having reached the mandatory retirement age of sixty-two, but still another attempt was to be made to install him as engineer in chief. In March, 1885 the first postwar Democratic administration took office, and within a few months the new Secretary of the Navy, William C. Whitney, received a letter from the famous shipbuilder, Nathaniel McKay, urging that Isherwood be brought out of retirement so that he might apply his talents and experience to designing the vessels of the new Navy. Both a good Democrat and a "consumate engineer," perhaps the greatest in the world, Isherwood had been shamefully treated by the series of Republican administrations since the Civil War, McKay explained, and now it was only just that he be rewarded and that the Navy profit by his appointment as engineer in chief.[11] Once a civilian, however, Isherwood was to remain one, although he would finally receive in retirement that which had been denied him on active duty: the rank of rear admiral.

Retirement, for Isherwood, did not mean relaxation. His loyal friend Theodore Zeller virtually moved into Isherwood's house, and until Zeller died in 1901 the two men would work all day in Isherwood's massive bedroom, which served as both library and study, collecting experimental data, often from foreign sources, which would then be organized and published as lengthy articles in scientific journals over the next quarter-century.

Early in 1872, after his return from Mare Island, Isherwood had moved his mother, wife, and children into a five-story, rust-red sandstone house at 111 East 36th Street. It was next door to Admiral David Farragut's home, and a few doors east of Park Avenue in the fashionable Murray Hill area of New York city. From their financial transactions during the Civil War and afterwards, Isherwood and his mother had accumulated a sizeable fortune; their house was a measure of their wealth. Solid mahogany and sterling silver fixtures were everywhere, even in Isherwood's private bathroom, where stood a huge copper bathtub, reputed to be one of the first installed in a private home in New York city. The rest of the house matched this sumptuous decor, for Isherwood brought back from his travels in Europe a great assortment of elaborate Second Empire furnishings. Hiring a number of skilled but indigent immigrant workmen during the depression following the panic of 1873, he had them decorate his home with frescoes, including a representation of Pompeii.[12]

Surrounded by his large family and a retinue of servants, Isherwood spent his last years in the Navy and those of retirement in unquestioned physical comfort, but all of this, including his family, was peripheral to his life of engineering. Refusing to join social clubs or most professional societies, never entertaining at home, treating card playing as a waste of the intellect, and considering most of the human race with whom he came in contact as stupid and annoying, Isherwood preferred the companionship of Theodore Zeller and his wife. In the evenings, Isherwood for years invariably left his own home to spend the few hours of leisure he permitted himself with the Zellers, either at their home or at the opera, of which he had remained an avid devotee.

By the 1880's engineering had become such a quickly expanding profession that the need for organization was apparent. As in so many other areas of American life at this time, there was an effort to bring together and share information and experiences and to develop a conscious group identity. In 1888 the American Society of Naval Engineers came into existence, and soon its members were busy preparing the publication of a professional journal. Departing from his usual custom, Isherwood became a member of the society, but he declined the position of editor of its journal, pleading age as well as his commitment to research in engineering as an excuse. Nevertheless, when the first issue of the society's journal appeared in February, 1889, Isherwood contributed the lead article, and for many years he continued to publish long articles in this magazine. Refusing to join other professional societies, in 1894 he was elected an honorary member of the American Society of Mechanical Engineers, an organization which would continue to honor him even after his death.[13]

By the late 1890's, Isherwood had outlived most of his contemporaries, to become a grand old man of engineering. Both for his Civil War contributions and for his extensive writings in engineering which had increased, rather than diminished, during his retirement, he received a series of honors and tributes which culminated in a testimonial dinner held at Philadelphia, in April, 1898. A number of friends and associates had subscribed funds during the previous year to commission a bronze bust of Isherwood which was then presented to him together with a series of colorful recollections and flowery tributes.[14]

Isherwood stated that this event marked his passage from active life, but for another ten years he would continue his research and writing before he finally began to fail. Losing most of his fortune in the "bankers" panic of 1907, Isherwood had to change his way of life drastically, but by this time his mother had been dead for ten years and his wife and the Zellers were also gone. Enfeebled by age, he was forced to curtail his writing, and his years after 1910 were spent as an invalid.

Although he had received the full measure of tribute from the engineering profession, it was not until his very last days when the Navy decided to recognize his accomplishments and contributions. By 1904, on the grounds of the United States Naval Academy, at Annapolis, a Steam Building had been constructed to house the departments of marine engineering and naval construction, and the Officers' Postgraduate School. On May 15, 1915, Isherwood received a letter from the academy stating that the Navy Department had formally approved that the building be called Isherwood Hall, that a copy of the bronze bust would be placed on the capstone over the main entrance, and that a bronze tablet inside the building would commemorate the beginning, during his administration as Engineer in Chief, of the school of marine engineering at the academy.

One month later, on June 19, 1915, in his ninety-third year, Benjamin Isherwood died. By this time both the United States Navy and the profession of engineering had advanced far beyond the mid-nineteenth century world in which he had played his most significant historical role. Yet his work had an unmistakable impact on the Navy and the art of engineering at the time of his death. As Engineer in Chief from 1861 to 1869, he had held a vital post during a time of crisis for the future of his nation and for his profession. By unremitting tenacity and ingenuity he had met the demands of his nation; by example and exhortation he had directed the growth of his profession, defending it from the evils, as he put it, of "stupid inertia, or actively jealous opposition . . . or crude notions of complacent folly." An iconoclast in an age of engineering idolatry, his motto could serve as his epitaph: "It is Truth we are after, nothing more and nothing less."[15]

# Notes

*(Written originally as a dissertation, this work was supported by highly detailed documentation, some of which the publisher has eliminated.)*

## Chapter I
### The New Engineer in Chief

[1] Roy P. Basler, ed., *The Collected Works of Abraham Lincoln,* IV, 297.

[2] *The New York Times,* March 29, 1861.

[3] Quoted in *The Celebration of the Centennial Anniversary of the Founding of the Albany Academy, May 24, 1913.*

[4] Gideon Hawley, *Address Delivered at the Public Exercises of the Albany Academy, in the Second Dutch Church, August 6, 1835.*

[5] Ernest J. Muller, quoted in *Celebration of the Seventy-Fifth Anniversary of the Founding of the Albany Academy, October 25, 1888.*

[6] *Statutes of the Albany Academy, Revised and Passed, October 9, 1829. Reprinted, with Amendments and Alterations, September, 1843.*

[7] Records of the Albany Academy for Benjamin F. Isherwood.

[8] W. K. Latimer to Isherwood, December 24, 1845, "Testimonial Letters to Chief Engineers," The National Archives, Record Group 24.

[9] Frank M. Bennett, *The Steam Navy of the United States* (hereafter referred to as *Steam Navy*), 82.

[10] According to "Abstracts of Service Records, etc.," The National Archives, Record Group 24, Isherwood's date of rank was later adjusted retroactively to October 31, 1848, after a successful petition to Congress in which he complained that by being overseas he had once again been unable to take his examination for promotion with the rest of the first assistant engineers and, therefore, had been penalized by being assigned a later date of rank.

[11] 33 Cong., 1 sess., *House Exec. Doc. No. 65,* 155.

[12] Bennett, *Steam Navy,* 56.

[13] Charles B. Stuart, *The Naval and Mail Steamers of the United States,* 105.

[14] "Memorial of Engineers of the Navy Praying a Reorganization of the Corps to Which They Belong," 32 Cong., 1 sess., *Sen. Misc. Doc. No. 45,* 3.

[15] R. W. Meade, II manuscripts, The New-York Historical Society.

[16] William Maxwell Wood, *Fankwei: Or, the San Jacinto in the Seas of India, China and Japan,* 265.

[17] *Ibid.,* 274.

[18] "Logs, Journals, and Diaries of Officers of the United States Navy at Sea, No. 87; Medical Journal of Assistant Surgeon R. P. Daniel aboard the USS *San Jacinto,* on a Cruise from New York to Hongkong and Back to New York, October 25, 1855–August, 1858," The National Archives, Record Group 45.

[19] *Ibid.*

[20] *Ibid.*

[21] "Surrender and Destruction of Navy Yards . . . ," 37 Cong., 2 sess., *Senate Report No. 37*, 2–3, 87; Bennett, *Steam Navy*, 230–31; Gideon Welles, *The Diary of Gideon Welles* (hereafter referred to as *Diary*), I, 39–42. The *Merrimack*, valued at $1,200,000 fully equipped, was worth more than all the other vessels in the Norfolk yard combined.

[22] "Report of the Secretary of the Navy," 37 Cong., 1 sess., *Sen. Exec. Doc. No. 1*, 87; Bennett, *Steam Navy*, 232, 236; Welles, *Diary*, I, 43.

[23] *Official Records of the Union and Confederate Navies in the War of the Rebellion* (hereafter referred to as *Official Records of the Navies*), Ser. 1, Vol. IV, 276, 290–91.

[24] *Ibid.*, 281.

[25] Bennett, *Steam Navy*, 239.

[26] *Ibid.*, 239–40.

[27] *Ibid.*, 241; Interview with Mrs. Madeleine Kerwin, November 12, 1962.

## Chapter II
### Building the Union Navy

[1] 42 Cong., 2 sess., *House Misc. Doc. No. 201*, 294.

[2] Of the ninety vessels in the Navy, thirty-four were steamers, of which twenty-six were in commission.

[3] As Isherwood saw it, the key to victory in the Civil War lay in this prevention of blockade-running and the disruption of the Confederate economy. "After exhaustion had arrived," he later recalled, "any strategy was effective, any generalship was successful, and all battles were victories." B. F. Isherwood, "The Sloop-of-war 'Wampanoag,'" *Cassier's Magazine*, Vol. XVIII (August, 1900), 282–84.

[4] R. H. Thurston, "Benjamin F. Isherwood," *Cassier's Magazine*, Vol. XVIII (August, 1900), 345.

[5] *Diary*, I, 74.

[6] William Howard Russell, *My Diary North and South*, II, 301.

[7] "Dedication of a Bronze Bust of Chief Engineer Isherwood, U. S. Navy," *Journal of the American Society of Naval Engineers* (hereafter referred to as ASNE *Journal*), Vol. X (May, 1898), 466.

[8] *Ibid.*, 459–60.

[9] These engines were of the horizontal, back-acting type, with modifications by Isherwood in the connecting of the piston rods to the crankshaft. The engines had 2 cylinders of 30-inch diameter and 18-inch stroke, with a cutoff set at 2/3 of the stroke. Two Martin water-tube boilers and a Sewell surface condenser were used, and blowers for forced draft were added later. The engines operated under a normal maximum boiler pressure of 30 pounds per square inch above atmospheric pressure. At the cruising speed of 6.8 knots, and developing an indicated horsepower (IHP) of 200, the vessel used 3.2 pounds of coal per indicated horsepower per hour, a fairly inefficient result from engine operation, according to Thomas Main, in *The Progress of Marine Engineering: From the Time of Watt until the Present Day* (hereafter referred to as *The Progress of Marine Engineering*), 189.

[10] According to John W. Oliver, in his *History of American Technology*,

287, the success of the southern blockade-runners forced naval engineers to make a number of experiments, including the introduction of the forced-draft system in boilers, in an attempt to increase the speed of Union warships.

The single cylinder of this engine was forty-four inches in diameter and had a seven-foot stroke, with the cutoff set at two-thirds. Isherwood used two of the Martin boilers with steam pressure at thirty pounds. As with his other engines, the efficiency of this one was low, burning three to three and one-half pounds of coal per IHP hour.

[11] To obtain greater speed, Isherwood enlarged his single inclined engine. The cylinder measured fifty-eight inches in diameter and had a stroke of eight feet, nine inches. The cutoff, boilers, and efficiency, as measured by coal consumption, were the same as for the previous class. Main, *The Progress of Marine Engineering*, 193.

[12] Isherwood used a pair of horizontal, back-acting engines for each of these ships. The cylinders measured forty-two inches in diameter with a thirty-inch stroke, the cutoff set at two-thirds. With two Martin boilers supplying steam at thirty pounds pressure, the engines developed up to nine hundred IHP, and burned coal at a rate of three to three and one-half pounds per IHP hour. Main, *The Progress of Marine Engineering*, 191–92.

[13] Only three vessels of this class had Isherwood engines. Of the other steamers, one utilized English engines, one had a segmental engine of William Wright's design, one had a direct-acting engine built by Merrick, and two had vibrating-lever engines, designed by Ericsson and Corliss. 38 Cong., 2 sess., *House Report No. 8*, 7–9.

[14] "Report of the Secretary of the Navy," 38 Cong., 2 sess., *House Exec. Doc. No. 1*, 1094; Thurston, "Benjamin F. Isherwood," *Cassier's Magazine*, Vol. XVIII (August, 1900), 344.

[15] Piston rods on Isherwood's ninety-day gunboats were 2.78 times heavier than they would ordinarily be for engines of that size. According to Thomas Main, in *The Progress of Steam Engineering*, 189, the other engine parts were in the same proportion, and this fact held for all classes of warships. Most engineers of that day, including Main, believed these unusually heavy engine parts to be much stronger than necessary, and they deplored the uneconomic results Isherwood naturally obtained in his efforts to assure mechanical durability.

[16] Farragut to Welles, July 30, 1863, *Official Records of the Navies*, Ser. 1, Vol. XX, 428–29.

[17] Welles to Isherwood, August 12, 1863. *Official Records of the Navies*, Ser. 1, Vol. XX, 456.

[18] By 1864 the Navy had suffered such damage from poor materials and bad workmanship in the construction of its vessels that Welles, in January, sent out a printed circular to all inspectors strongly encouraging them to bear down on the private contractors, and to "inexorably extract" the highest degree of workmanship from them.

[19] Isherwood's testimony of April 18, 1872, 42 Cong., 2 sess., *House Misc. Doc. No. 201*, 206–207.

[20] "Letters from Bureaus," The National Archives, Record Group 45.

[21] Isherwood to Fox, August 10, 1862, "Letters from Bureaus," The National Archives, Record Group 45.

[22] Isherwood to Fox, August 5, 1862, "Letters to Bureaus," The National Archives, Record Group 45.
[23] Isherwood to Welles, July 7, 1863, "Letters Sent to the Secretary of the Navy," The National Archives, Records of the Bureau of Ships (hereafter referred to as Record Group 19).
[24] Isherwood to Fox, June 28, 1863, "Letters from Bureaus," The National Archives, Record Group 45.
[25] Isherwood to Welles, February 27, 1864, "Letters Sent to the Secretary of the Navy," The National Archives, Record Group 19.
[26] "Letters Sent to the Secretary of the Navy," The National Archives, Record Group 19.
[27] "Report of the Secretary of the Navy," 39 Cong., 2 sess., *House Exec. Doc. No. 1*, 177–78.
[28] "Report of the Secretary of the Navy," 40 Cong., 2 sess., *House Exec. Doc. No. 1*, 181.
[29] Isherwood to Secretary of the Navy Robeson, January 18, 1875, "Letters from Officers of Rank below That of Commander," The National Archives, Record Group 45.
[30] "Iron-Clad Ships, Ordnance, etc.," 37 Cong., 2 sess., *House Misc. Doc. No. 82*, 1; "Report of the Secretary of the Navy," 38 Cong., 2 sess., *House Exec. Doc. No. 1*, xxviii–xxix.
[31] Isherwood to Welles, April 29, 1863, "Letters from Bureaus," The National Archives, Record Group 45.
[32] *Senate Executive Journal*, XIII, 5.
[33] Gideon Welles to Edgar Welles, January 25, 1863, Gideon Welles Papers, The Library of Congress.
[34] Gustavus Vasa Fox Papers (hereafter referred to as Fox Papers), The New-York Historical Society.
[35] *Ibid*.
[36] Isherwood to Secretary of the Navy George Robeson, January 18, 1875, "Letters from Officers of Rank below That of Commander," The National Archives, Record Group 45.

# Chapter III

## *Isherwood and the Ironclads*

[1] The most successful of Civil War ironclads was not a monitor-type vessel, but rather the ironclad casemate *New Ironsides,* begun in 1861 and completed in late 1862, by Merrick and Sons, of Philadelphia. This vessel carried 4 inches of armor over a white-oak hull and displaced 4,120 tons. With machinery delivering only 700 horsepower, she could barely attain a speed of 6 knots, and thus was impractical for sea duty. However, with her heavy battery of sixteen 11-inch Dahlgren smoothbore cannon, two 200-pounder Parrott rifles, and four 24-pounder howitzers, the *New Ironsides* was an extremely formidable vessel for naval operations along the coastline and in bays and river mouths. During the Civil War this ironclad had a remarkable record, and was reputed to have been in action more days than any other vessel of the Union Navy. Also possessing the record of having

been hit by enemy fire more times than any other vessel, the *New Ironsides* remained operational and virtually invulnerable throughout the war.

The *New Ironsides* might have greatly changed the subsequent direction of American warship design had she been finished six months earlier and gone to Hampton Roads to meet the *Virginia*. As it was, the *Monitor* was the first of the American ironclads in action, and her stalemate with the *Virginia* proved sufficient to assure the continuation of the monitor principle in American naval construction and strategy, while overseas the major naval powers built their warships along the lines of the *New Ironsides* design. Bennett, *Steam Navy*, 272–74.

[2] James Phinney Baxter, III, *The Introduction of the Ironclad Warship*, 270–75. Fox to Welles, August 15, 1863, Welles Papers, The Library of Congress.

[3] Fox to Welles, August 15, 1863, Welles Papers, The Library of Congress.

[4] Fox Papers, The New-York Historical Society.

[5] "Light-Draught Monitors," *Report of the Joint Committee on the Conduct of the War* (hereafter referred to as "Light-Draught Monitors"), Part III, 38 Cong., 2 sess., *Sen. Report No. 142*, 111.

[6] *Ibid.*

[7] *Ibid.*

[8] *Ibid.*, 111–12.

[9] *Ibid.*

[10] *Ibid.*

[11] *Ibid.* The British naval expert Sir Thomas Brassey was so taken with this passage that he quoted it in the introduction to volume three of his *The British Navy: Its Strength, Resources, and Administration*. Brassey argued that Britain had a particular need for an ocean-going fleet to keep open its lines of communication with its colonies. The British Navy could not afford, Brassey recognized, to be merely a passive, defensive force for its own coasts and harbors.

[12] 37 Cong., 3 sess., *House Exec. Doc. No. 1*, 32–35.

[13] 38 Cong., 1 sess., *House Exec. Doc. No. 1*, xiii.

[14] "Letters to the Secretary of the Navy," The National Archives, Record Group 19.

[15] Smith to Welles, May 10, 1862, "Letters from Bureaus of the Navy Department," The National Archives, Record Group 45.

[16] Isherwood, Hartt, and Martin to Welles, May 13, 1862, "Letters from Bureaus of the Navy Department," The National Archives, Record Group 45.

[17] "Ship Construction," The National Archives, Record Group 45.

[18] "Charges against the Navy Department," 42 Cong., 2 sess., *House Misc. Doc. No. 201*, 394; *Army and Navy Journal*, October 23, 1869, 144–45.

[19] Eads to Fox, February 1, 1864, and Eads to Lenthall, February 2, 1864, "Ship Designs," The National Archives, Record Group 45.

[20] 38 Cong., 2 sess., *Congressional Globe*, 866–67. Although Grimes did not become chairman of the Senate Naval Affairs Committee until 1865, Welles and the Navy Department relied heavily on his support. The incumbent chairman, John P. Hale, had become an inveterate opponent of the department, and he was the last man Welles could look to for congressional aid.

[21] Welles, *Diary*, II, 350.

[22] "Light-Draught Monitors," 83.

[23] Charles H. Cramp, quoted in Augustus C. Buell, *The Memoirs of Charles H. Cramp*, 80, 87.
[24] Ericsson to Fox, December 22, 1862, Fox Papers, The New-York Historical Society.
[25] Lenthall to Welles, December 23, 1864, "Letters from Bureaus," The National Archives, Record Group 45; "Light-Draught Monitors," i–ii.
[26] Ericsson to Fox, February 24, 1863, Fox Papers, The New-York Historical Society.
[27] Gregory to Fox, June 27, 1863, Fox Papers, The New-York Historical Society.
[28] Gregory to Fox, June 15, 1863, Fox Papers, The New-York Historical Society.
[29] Stimers to Fox, August 27, 1863, Fox Papers, The New-York Historical Society.
[30] Stimers to Fox, February 18, 1864, Fox Papers, The New-York Historical Society.
[31] Fox Papers, The New-York Historical Society.
[32] Stimers to Fox, June 8, 1864, Fox Papers, The New-York Historical Society.
[33] Welles to Fox, June 10, 1864, Fox Papers, The New-York Historical Society.
[34] Stimers to Fox, June 16, 1864, Fox Papers, The New-York Historical Society.
[35] Fox Papers, The New-York Historical Society.
[36] Fox to Ericsson, July 28, 1864, Fox Papers, The New-York Historical Society.
[37] Welles, *Diary*, II, 81–82, 108.
[38] *Ibid.*, 350.
[39] *Ibid.*, 226.
[40] Stimers' testimony, "Light-Draught Monitors," 93–94, 103.
[41] Isherwood's testimony, "Light-Draught Monitors," 116.
[42] *Ibid.*, 115–17.
[43] *Ibid.*, 120.
[44] Welles, *Diary*, II, 351.
[45] *Ibid.*, I, 401; II, 232–33.
[46] "Light-Draught Monitors," 119.

# Chapter IV

## *The Unwelcome Pioneer*

[1] Edward Cressy, *A Hundred Years of Mechanical Engineering*, 35–37.
[2] *Cassier's Magazine*, Vol. XVIII (August, 1900), 282.
[3] According to Walter M. McFarland, in the ASNE *Journal*, Vol. XXVIII, 2, this book was the first in which there appeared diagrams of indicator cards actually taken from working marine engines. Other books presented ideal diagrams drawn on the basis of theoretical results, while Isherwood's book was unique in utilizing diagrams taken from practical operation.
[4] *Engineering Precedents*, II, vii.

[5] *Ibid.*, ix–x.

[6] In the tenth edition, published in 1865, of their book, *Lessons and Practical Notes on Steam, the Steam Engine, Propellers* . . . , W. H. King and Chief Engineer J. W. King, U. S. Navy, explained why they used Mariotte's law: "This theory would be literally correct did the temperatures remain constant; but as the temperature of all gases becomes reduced by expansion, the law does not hold good; nevertheless, in the steam engine, where there are so many extraneous circumstances which practically affect all calculations appertaining to the same, it is considered all that is ever required, and from its extreme simplicity is universally adopted." According to Bennett, *Steam Navy*, 610, King's textbook for many years was regarded as the best American work of its kind, and by going through nineteen editions, more copies were sold than of any other book written by a naval officer up to the time of Mahan.

[7] *Engineering Precedents*, II, 48.

[8] *Ibid.*, 50.

[9] R. H. Thurston, in *Development of the Philosophy of the Steam Engine*, 24, estimated that in steam engines with unjacketed cylinders, the loss of efficiency from condensation was rarely less than 25 per cent and was often more than equal to the entire quantity of heat transformed into mechanical energy.

[10] *Engineering Precedents*, II, 76.

[11] *Journal of the Franklin Institute*, Vol. LXIX (April, 1860), 284.

[12] *Engineering Precedents*, II, v.

[13] *Ibid.*, viii.

[14] Isherwood, *Experimental Researches in Steam Engineering* (hereafter referred to as *Experimental Researches*), I, 91, 100, 119, 138–39.

[15] In addition, the compound engine, which utilized a small, high-pressure cylinder feeding its exhausted steam into a larger low-pressure cylinder, was in use well before the time that Isherwood experimented on simple engines. However, the compound engine was still in its experimental stages and was not practical for warship propulsion. Although the compound engine had successful application in 1854 through the work of John Elder, and was in general use on certain merchant lines during the 1860's, it was not used in transatlantic service nor in the British Navy until the 1870's, and even then there was considerable resistance to its use in the Royal Navy. Similarly, the compound engine did not come into use in the American Navy until the early 1870's. The difficulty with this engine was not in lack of economy in operation, but rather in its unreliability and high cost of maintenance. Edgar C. Smith, *A Short History of Naval and Marine Engineering*, 174–81.

Isherwood, concerned with the practical operations of engines generally in use during the 1860's, paid little attention in his writings to the compound engine until later years. However, his theories on the economical limits of steam expansion contributed to the later development of efficient compound-engine propulsion.

[16] *Experimental Researches*, I, 120.

[17] Page 281.

[18] Thurston, *Manual of the Steam Engine*, 276; Thurston, *The Development of the Steam Engine: An Historical Sketch*, 34–35.

[19] Isherwood's report of October 12, 1865, contained in the "Report of the Secretary of the Navy," 39 Cong., 1 sess., *House Exec. Doc. No. 1*, 306–307.

[20] Historians of technology claim that Isherwood coined the term "atomize" through his work on using oil as an engine fuel. Hot oil would be sprayed, or "atomized," with superheated steam and the resultant mist would be ignited to provide heat. Charles Singer, *et al.*, eds., *A History of Technology, V: The Late Nineteenth Century, c. 1850 to c. 1900*, 117.

[21] Isherwood to Welles, February 7, 1863, "Letters from Bureaus," The National Archives, Record Group 45.

[22] Many engineers apparently agreed, since over fifty years later it was still of very great value, and the book at that time was praised as "a store house of information to which for a long time there was no parallel." *Army and Navy Journal*, July 3, 1915, 1390.

[23] *Experimental Researches*, I, xv.

[24] Isherwood to Welles, May 20, 1864, "Letters Sent to the Secretary of the Navy," The National Archives, Record Group 19.

[25] Endorsement by Chief Clerk William Faxon, May 23, 1864, on the letter from Isherwood to Welles, May 20, 1864, "Letters Sent to the Secretary of the Navy," The National Archives, Record Group 19.

[26] *Experimental Researches*, II, xv, xxi, xcii.

[27] Page 409.

[28] Thurston, *Cassier's Magazine*, Vol. XVIII (August, 1900), 346; Thurston's remarks in the ASNE *Journal*, Vol. X, 443.

[29] Although Isherwood's methods and principles of experimentation are so commonplace today that they may not seem especially striking to the modern reader, one must always remember the rudimentary state of technology in the mid-nineteenth century and the complete lack of professional training for engineers. It was Isherwood's work in these circumstances that inspired Dr. Hollis, president of the American Society of Mechanical Engineers, to state in an article in *Power* magazine, Vol. XLVII (January 15, 1918), 104, "I think of him as perhaps the father of our great research laboratories in engineering, as his investigations in connection with steam engines and with boilers preceded all of our schools of mechanical engineering."

[30] *Cassier's Magazine*, Vol. XVIII (August, 1900), 284.

[31] *Ibid.*, 284–85.

[32] *Experimental Researches*, I, xiv.

[33] *Ibid.*, I, xxvi. An avid opera-lover, Isherwood metaphorically described this ability to discern scientific principles as the "nice perception that detects the ever present *tema* amid the infinite variations which compose the grand opera of nature."

[34] *Experimental Researches*, I, xxv, 139–40; II, xviii.

[35] *Ibid.*, I, xxv.

[36] *Ibid.*, I, xxvi.

[37] *Ibid.*, I, xxv.

[38] *Ibid.*, II, xvi–xvii.

[39] *Ibid.*, II, xvi.

[40] *Ibid.*, I, xiv.

[41] *Ibid.*, I, xv, xxiv–xxv; II, xxii–xxiii.

[42] *Ibid.*, II, xxii–xxiii.

[43] *Ibid.*, II, xxii.

[44] *Ibid.*, II, xvii–xviii.

## Chapter V

### The Lawyer and the Engineer

[1] H. W. Howard Knott, "Edward Nicoll Dickerson," *Dictionary of American Biography*, V, 288–89. This general account of the *Pensacola*'s construction is by Isherwood in an official report to Welles, 37 Cong., 3 sess., *Sen. Exec. Doc. No. 45*, 4–8.

[2] *Ibid.*, 6.

[3] *Ibid.*, 10.

[4] Dickerson to Welles, January 6, 1863, Welles Papers, The Library of Congress.

[5] Dickerson, *The Steam Navy of the United States: Its Past, Present and Future* (hereafter referred to as *Navy of the United States*), 1–3, 14–17.

[6] *Ibid.*, 15.

[7] *Ibid.*, 20.

[8] *Ibid.*, 66.

[9] *Ibid.*, 3.

[10] *Ibid.*, 4.

[11] *Ibid.*, 43–44.

[12] *Ibid.*, 77.

[13] *Ibid.*

[14] *Ibid.*, 80, footnote.

[15] *Ibid.*, 6, 9, 29.

[16] Welles, *Diary*, I, 504–505.

[17] Gideon Welles to Edgar Welles, January 17, 1864, Welles Papers, The Library of Congress.

[18] Page 281.

[19] Bennett, *Steam Navy*, 165–66.

[20] *The New York Times*, January 8, 1864.

[21] *Ibid.*, July 28, 1864.

[22] *Ibid.*, May 28, 1864; June 4, 1864.

[23] Forbes to Fox, June 9, 1863, Fox Papers, The New-York Historical Society.

[24] Ralph R. Gurley, "The *Wampanoag*," *United States Naval Institute Proceedings*, Vol. LXIII (December, 1937), 1733. Welles later insisted, in a letter to Representative A. H. Rice, January 22, 1866, that Forbes virtually forced the contract on the unwilling Navy Department, which had to accept it to avoid giving justification to the complaints that the department was prejudiced against all private engineers and designers. "Letters to Congress," The National Archives, Record Group 45.

[25] George W. Quintard to Gustavus Fox, August 1, 1864, Fox Papers, The New-York Historical Society. (It encloses a copy of a letter from Forbes to Quintard, July 29, 1864.)

[26] John Griswold to John Ericsson, December 8, 1864, Ericsson Papers, The New-York Historical Society.

## Chapter VI
### Trials and Tribulations

[1] *Steam Navy*, 518.

[2] As recorded in the *Journal of the Franklin Institute*, Vol. LXXXI (March, 1866), 206, Forbes agreed to the following conditions:
   1. Materials, workmanship, detail, and finish of the machinery should be first class.
   2. Machinery performance should show the strength, reliability, practical efficiency, and durability of all components.
   3. The contractor must not exceed the weight of the Government machinery, increase its space, or decrease the amount of coal bunkered in the vessel.
   4. If, after the trial, the *Algonquin*'s performance was inferior to the *Winooski*'s, either in amount of power developed or in the cost of the power in coal, Forbes would remove his own machinery and replace it at his own cost with that designed by Isherwood for the *Winooski*.

[3] *The New York Times*, August 3, 1865. A letter to the editor signed "Justitia." Internal evidence strongly indicates that it was written either by Isherwood or one of his associates.

[4] Dickerson, as quoted in Bennett's *Steam Navy*, 519.

[5] *Diary*, II, 346.

[6] *Ibid.*, II, 356, 361.

[7] Forbes to Fox, July 11, 1865, and August 22, 1865, Fox Papers, The New-York Historical Society.

[8] Isherwood to Welles, August 15, 1865, "Letters from Bureaus of the Navy Department," The National Archives, Record Group 45.

[9] *The New York Times*, August 25, 1865.

[10] *Ibid.*, August 25 and 26, 1865.

[11] Isherwood to Welles, August 29, 1865, Welles Papers, The Library of Congress.

[12] Gregory to Fox, September 10, 1865, Fox Papers, The New-York Historical Society.

[13] Welles to Fox, September 13, 1865, Fox Papers, The New-York Historical Society.

[14] Isherwood to Fox, September 13, 1865, Fox Papers, The New-York Historical Society.

[15] Isherwood to Fox, September 21, 1865, Fox Papers, The New-York Historical Society.

[16] *The New York Times*, September 23, 1865.

[17] *Ibid.*, September 25, 1865.

[18] *Ibid.*

[19] *Ibid.*, September 26, 1865.

[20] Gregory to Fox, September 26, 1865, Fox Papers, The New-York Historical Society.

[21] *Army and Navy Journal*, September 30, 1865, 85–86; October 7, 1865, 106; October 14, 1865, 122–23. The London *Engineer* took on a patronizing

air and denounced both participants impartially, branding them "equally ignorant of the fundamental principles of the subject on which they presume to discourse." October 21, 1865, 255.

[22] Ericsson to Fox, September 30, 1865, Ericsson Papers, The New-York Historical Society.

[23] Isherwood to Fox, October 6, 1865, Fox Papers, The New-York Historical Society.

[24] Forbes to Gregory, quoted in Bennett, *Steam Navy*, 522.

[25] Page 119.

[26] *Journal of the Franklin Institute*, Vol. LXXXI, 202; *The New York Times*, February 16, 1866.

[27] *Journal of the Franklin Institute*, Vol. LXXXI, 203–209; Isherwood to Fox, February 16, 1866, "Letters from Bureaus," The National Archives, Record Group 45.

[28] *Journal of the Franklin Institute*, Vol. LXXXI, 207–208.

[29] Bennett, *Steam Navy*, 523.

[30] *Engineering*, March 23, 1866, 183.

[31] Bennett, *Steam Navy*, 523–24. According to Bennett, Chief Engineer E. D. Robie, on board the *Idaho* as an observer, made a thorough examination of the machinery before its trial, and then quietly insured his life for a large sum, for the benefit of his family, before obediently returning to perform his duty in the face of hazard.

[32] "Letter from the Secretary of the Navy . . . relative to the trial of the United States Steamer *Wampanoag*, and vessels of that class . . . ," 40 Cong., 2 sess., *House Exec. Doc. No. 339*, 45–47; Bennett, *Steam Navy*, 524–26.

[33] Isherwood to Fox, April 21, 1866, Fox Papers, The New-York Historical Society.

[34] Isherwood to Fox, April 24, 1866, Fox Papers, The New-York Historical Society.

[35] Bennett, *Steam Navy*, 529.

# Chapter VII

## *The Steam Bureau under Attack*

[1] 38 Cong., 1 sess., *House Exec. Doc. No. 1*, xxxv; 37 Cong., 1 sess., *House Exec. Doc. No. 1*, 918.

[2] *The New York Times*, February 1, 1863.

[3] Welles to Isherwood, January 21 and 28, 1863 and February 9, 1863, "Letters to Bureaus," The National Archives, Record Group 45.

[4] "Charges against the Navy Department," 42 Cong., 2 sess., *House Misc. Doc. No. 201*, 390–401. This investigation of 1872 contains documents of the board to examine Isherwood steam machinery.

[5] Welles, *Diary*, I, 519.

[6] *Ibid.*, I, 528–29. For a colorful summary of Welles's attitude toward Hale, see *Ibid.*, II, 52.

[7] "Select Committee on Naval Supplies," 38 Cong., 1 sess., *Sen. Report No. 99, passim*. On February 17, 1865, Senator James R. Doolittle, of Wisconsin, defended Isherwood against Hale's attacks in Congress by discussing the testimony and findings of the Select Committee on Naval Supplies which

showed, Doolittle maintained, that Isherwood's official integrity was never compromised in bureau transactions. 38 Cong., 2 sess., *Congressional Globe,* 856–57.

[8] Welles, *Diary,* II, 87.

[9] *Ibid.,* 103–105.

[10] 38 Cong., 2 sess., *Congressional Globe* (February 3, 1865), Appendix, 35.

[11] 38 Cong., 1 sess., *Congressional Globe* (February 1, 1864), 422–23.

[12] "Marine Engines," 38 Cong., 2 sess., *House Report No. 8,* 1.

[13] *Ibid.,* 12.

[14] *Ibid.,* 1–2.

[15] *Ibid.,* 12–14.

[16] *Ibid.,* 32.

[17] *Ibid.*

## Chapter VIII

### *The Inventor and the Engineer*

[1] Donald Nevius Bigelow, *William Conant Church and the Army and Navy Journal,* 198.

[2] *Ibid.,* 208, 210–11, 214–15, 221. Church's undeviating loyalty to Ericsson culminated in his lengthy and eulogistic biography of the inventor, published in 1890.

[3] Ericsson to Fox, July 28, 1862, John Ericsson Papers (hereafter referred to as Ericsson Papers), The Library of Congress; Ericsson to Fox, August 15, 1863, John Ericsson Papers (hereafter referred to as Ericsson Papers), The New-York Historical Society.

[4] Ericsson to Fox, September 26, 1863, Fox Papers, The New-York Historical Society.

[5] A slightly larger ship, the *Pompanoosuc,* was also begun by the Navy Department, but was never completed. The proclivity of the Navy Department under Gideon Welles's administration to name its warships after Indian tribes or after rivers and regions which, in particular, bore New England Indian names, caused great exasperation among sailors and officers alike in the American Navy. Of Isherwood's cruisers, the *Wampanoag* had been named for a Massachusetts Indian tribe largely wiped out in King Philip's War, in 1676, and the *Ammonoosuc* came from a river in New Hampshire. This authentically American flavor, however, was lost on Navy men who soon rechristened their vessels with such derisive nicknames as "My-aunt-don't-know-me" (*Miantonomah*), "Pompey's-new-suck" (*Pompanoosuc*), or "Am-I-a-new-sucker" (*Ammonoosuc*). Even when avoiding such corruptions, sailors often mispronounced the cumbersome Indian names through an inability to grapple successfully with the strange sounds and syllables. Moreover, there was rarely a complete agreement over the spelling of many of these Indian names. *Madawaska* was commonly written as *Madawasca,* and *Algonquin* often appeared as *Algonkin*. The wholesale renaming of warships in late 1869, during President Grant's administration, was greeted with relief by many Navy men who

would have agreed wholeheartedly with Alfred T. Mahan who decided that the names of Isherwood's cruisers had been "extracted by a fine-tooth comb drawn through the tangle of Indian nomenclature."

[6] Ericsson to B. F. Delano, October 12, 1863, Ericsson Papers, The New-York Historical Society.

[7] A large number of these letters are in the Ericsson Papers, The New-York Historical Society.

[8] Griswold to Ericsson, February 20, 1864, Ericsson Papers, The New-York Historical Society.

[9] Ericsson Papers, The New-York Historical Society.

[10] Ericsson to Welles, November 27, 1865, Ericsson Papers, The New-York Historical Society.

[11] Griswold to Ericsson, April 3, 1866, Ericsson Papers, The New-York Historical Society.

[12] Ericsson to Griswold, April 4, 1866, Ericsson Papers, The New-York Historical Society.

[13] Welles to Ericsson, April 21, 1866, Ericsson Papers, The New-York Historical Society.

[14] Ericsson to Griswold, April 23, 1866, Ericsson Papers, The New-York Historical Society.

[15] Ericsson to Fox, April 23, 1866, Fox Papers, The New-York Historical Society.

[16] Ericsson to Griswold, May 14, 1866. Aaron Vail Collection, The New-York Historical Society.

[17] Ericsson to Bennett Woodcroft, May 15, 1866, Ericsson Papers, The New-York Historical Society. Comments of a similar intensity are in a letter from Ericsson to Griswold, May 25, 1866, Ericsson Papers, The New-York Historical Society.

[18] Ericsson to Griswold, June 13, 1866, Ericsson Papers, The New-York Historical Society.

[19] Ericsson to Gregory, June 16, 1866, Ericsson Papers, The New-York Historical Society.

[20] Ericsson to Griswold, June 20, 1866, Ericsson Papers, The New-York Historical Society.

[21] Isherwood to Gregory, June 23, 1866, Ericsson Papers, The New-York Historical Society.

[22] Ericsson Papers, The Library of Congress.

[23] Griswold to Ericsson, July 15, 1866, Ericsson Papers, The New-York Historical Society.

[24] Griswold to Ericsson, July 23, 1866, Ericsson Papers, The New-York Historical Society.

[25] Ericsson to Griswold, August 25, 1866, Ericsson Papers, The New-York Historical Society.

[26] Isherwood to Gregory, September 7, 1866, and Gregory to Ericsson, September 8, 1866, Ericsson Papers, The New-York Historical Society.

[27] Ericsson to Gregory, September 18, 1866, Ericsson Papers, The New-York Historical Society.

[28] Ericsson to Welles, September 17, 1866, and Ericsson to Captain Axel Adlersparre, in Stockholm, October 12, 1866, Ericsson Papers, The New-York Historical Society.

[29] Ericsson to Griswold, November 16, 1866, Ericsson Papers, The New-York Historical Society.

[30] Isherwood to Ericsson, November 17, 1866, Ericsson Papers, The New-York Historical Society.

[31] Ericsson Papers, The New-York Historical Society.

[32] Isherwood to Fox, December 22, 1866, Fox Papers, The New-York Historical Society.

[33] *Army and Navy Journal,* November 24, 1866, 220.

[34] *The Engineer,* December 7, 1866, as quoted in the *Army and Navy Journal,* December 29, 1866, 302.

[35] "Steam Log of the *Madawaska,* January 16, 1867," Ericsson Papers, The New-York Historical Society; Bennett, *Steam Navy,* 532–33. The low steam pressure was probably caused by Ericsson's fear of using all the steam generated by the boiler system which he had insisted was too large for his engines.

[36] "Letter from the Secretary of the Navy . . . ," 40 Cong., 2 sess., *House Exec. Doc. No. 339,* 17–20 (Official report of the *Madawaska's* trial); Ericsson to Fox, January 28, 1867, Fox Papers, The New-York Historical Society.

[37] "Report of the Secretary of the Navy," 40 Cong., 3 sess., *House Exec. Doc. No. 1,* 121 (Isherwood's report of October 22, 1868). Isherwood maintained that the inability of the *Madawaska* to steam for any length of time was because of the lack of durability of the engine, causing the journals to overheat.

[38] 40 Cong., 2 sess., *House Exec. Doc. No. 339,* 20.

[39] *Ibid.,* 17.

[40] Isherwood's testimony of April 18, 1872, 42 Cong., 2 sess., *House Misc. Doc. No. 201,* 314.

[41] Ericsson to Adlersparre, February 5, 1867, Ericsson Papers, The New-York Historical Society.

[42] Griswold to Ericsson, March 10, 1867, Ericsson Papers, The New-York Historical Society.

## Chapter IX

### Isherwood's Masterpiece

[1] "Report of the Secretary of the Navy," 39 Cong., 1 sess., *House Exec. Doc. No. 1,* 309. Isherwood's report of October 12, 1865.

[2] "Letters from Bureaus of the Navy Department," The National Archives, Record Group 45. The first report was signed by Isherwood on December 9, 1865.

[3] Remarks of Senator James W. Grimes on the naval appropriation bill, 40 Cong., 2 sess., *Congressional Globe,* 2245.

[4] Isherwood's report of October 22, 1868, 40 Cong., 3 sess., *House Exec. Doc. No. 1,* 119.

[5] Porter's report of September 25, 1866, 39 Cong., 2 sess., *House Exec. Doc. No. 1,* 75.

[6] Isherwood to Fox, October 14, 1867, Fox Papers, The New-York Historical Society.

[7] Isherwood to Fox, May 1, 1867, Fox Papers, The New-York Historical Society.

[8] 40 Cong., 2 sess., *House Exec. Doc. No. 1*, 174–75.
[9] *The New York Times*, January 11 and January 22, 1866.
[10] Isherwood to Fox, May 1, 1867. Fox resigned as Assistant Secretary of the Navy in 1866, and most of Isherwood's correspondence with him dates from this time. Although no longer holding an official post in the Navy Department, Fox continued to take an active interest in naval affairs and continued to receive informative letters from officers still on active duty. Fox Papers, The New-York Historical Society.
[11] 38 Cong., 2 sess., *Congressional Globe*, 864.
[12] Isherwood to Fox, December 22, 1866, Fox Papers, The New-York Historical Society.
[13] Fox Papers, The New-York Historical Society.
[14] Isherwood to Fox, July 26, 1867, Fox Papers, The New-York Historical Society.
[15] Isherwood to Welles, March 18, 1867, "Letters Received from the Secretary of the Navy," The National Archives, Record Group 19.
[16] Farragut to Welles, April 8, 1867, "Letters Received from the Secretary of the Navy," The National Archives, Record Group 19.
[17] Welles to Isherwood, April 13, 1867, "Letters to Bureaus of the Navy Department," The National Archives, Record Group 45.
[18] Welles to Fox, April 25, 1867, Fox Papers, The New-York Historical Society.
[19] Delano to Fox, September 12, 1863, Fox Papers, The New-York Historical Society.
[20] William Hovgaard, *Modern History of Warships*, 166. In the 1860's, marine engines ran so slowly that to obtain the necessary propeller speed, gearing was used for multiplication rather than for reduction, as in modern turbine-driven ships. Almost all engines of this period, however, were direct acting; and by employing gears, Isherwood was considered by most engineers to be anachronistic.
[21] C. W. Dyson, "A Fifty Year Retrospect of Naval Marine Engineering," ASNE *Journal*, Vol. XXX (May, 1918), 258. According to Thomas Main, in *The Progress of Marine Engineering*, 196–98, critics, in later years, maintained that the wooden gears were failures, as they quickly ground down when the engines were used even briefly. See also Buell, *Cramp*, 92.
[22] Isherwood, on August 8, 1864, wrote an aggrieved letter to Welles, upset because half of the most important boiler braces of the *Wampanoag* had been omitted. He insisted that the superintending engineer had no discretion to permit such a variation of the plans, and that the drawings and specifications had to be followed to the letter, especially since this ship was "one of the largest and most important vessels built by the Department." "Letters Sent to the Secretary of the Navy," The National Archives, Record Group 19.
[23] "Letters from Bureaus," The National Archives, Record Group 45.
[24] *Army and Navy Journal*, December 17, 1864, 269, and December 24, 1864, 285.
[25] *Ibid.*, December 22, 1866, 277–78; December 29, 1866, 293–94; January 5, 1867, 309–10.
[26] *Ibid.*, August 17, 1867, 827.
[27] *Ibid.*, August 17, 1867, 827, and October 5, 1867, 105. The initial crew

of the *Wampanoag* on her commissioning included two petty officers, twenty-eight seamen, eight ordinary seamen, twenty-six landsmen, fourteen boys, forty-six Marines, sixty-seven firemen, and fifty-three coal heavers. The large engine-room crew was caused by the dock trials of the engines taking place at the time of the commissioning. "Log Book of the U.S.S. Wampanoag" (September 17, 1867 to March 2, 1868), The National Archives, Record Group 24.

[28] *Army and Navy Journal,* October 19, 1867, 140–41, and December 21, 1867, 284.

[29] *Ibid.,* November 2, 1867, 172–73.

[30] Page 266.

[31] *Army and Navy Journal,* February 1, 1868, 373, 380.

[32] 40 Cong., 2 sess., *House Exec. Doc. No. 339,* 2–3. (Nicholson's report to Welles, February 18, 1868); "Causes of the Reduction of American Tonnage," 41 Cong., 2 sess., *House Report No. 28,* 248.

[33] 40 Cong., 2 sess., *House Exec. Doc. No. 339,* 3. This tonnage measurement is apparently in English long tons of 2,240 pounds.

[34] Paullin, *United States Naval Institute Proceedings,* Vol. XXXIX, 745; Dyson, ASNE *Journal,* Vol. XXX, 258. On the basis of her design and performance, the *Wampanoag* was chosen by the Navy Department in the 1930's to be one of six historic vessels whose models would decorate the walls of the David W. Taylor Model Basin, at the Washington Navy Yard. These vessels were to typify stages of progress in propulsion and hull design of American vessels, along the lines of research and experimental work carried on at the Taylor Model Basin. Other vessels honored along with the *Wampanoag* were the clipper *Sovereign of the Seas* (1852), the battleship *Delaware* (1910), the heavy cruiser *Salt Lake City* (1929), the destroyer *Worden* (1935), and the Coast Guard cutter *Itasca* (1930).

In addition to her original hull design, the *Wampanoag* contained such "other novel features, many of which are considered noteworthy engineering developments of more recent years" as separate boilers, roller thrust bearings on the propeller shaft, and a balanced rudder. U. S. Navy Department, *The David W. Taylor Model Basin,* 4, 7.

[35] The *Wampanoag* was not fully loaded, since she did not carry her ammunition, which weighed over fifty tons.

[36] Fox Papers, The New-York Historical Society.

[37] Letter from Isherwood to Fox, March 5, 1868. Isherwood also summarized the accomplishments of the *Wampanoag* in his annual bureau report to Welles on October 22, 1868, concluding that the results of the trial furnished "a complete refutation to the many false accounts and misrepresentations which have been circulated about them." Fox Papers, The New-York Historical Society; 40 Cong., 3 sess., *House Exec. Doc. No. 1,* 121–24.

[38] Remarks of Representative Elihu B. Washburne, 40 Cong., 2 sess., *Congressional Globe,* 1423.

[39] *Army and Navy Journal,* March 7, 1868, 460. Church later decided this wind was actually a gale which had helped the cruiser along by at least one and one-half knots an hour. *Ibid.,* April 18, 1868, 557.

[40] *Ibid.,* March 7, 1868, 460; March 21, 1868, 485; and April 11, 1868, 533.

[41] *Engineering,* March 27, 1868, 270–71, 282.

[42] *The Engineer,* April 24, 1868.

[43] *Army and Navy Gazette,* March 13, 1868, as quoted in the *Journal of the Franklin Institute,* Vol. LXXXV (May, 1868), 327.

[44] According to the British naval expert Fred T. Jane, the *Inconstant* marked, for the first time in the British Navy, the theory that speed and gun power, "the offensive," might be profitably exploited at the cost of defensive strength. This theory later found full expression in the battle cruiser design of World War I. Accordingly, Jane stated, the *Wampanoag* inspired the design of the *Inconstant,* which, in turn, "represents the germ idea of our present battle-cruisers, and is supremely important on that account." *The British Battle Fleet,* I, 320–21.

[45] 40 Cong., 2 sess., *House Exec. Doc. No. 339,* 7.

[46] *Ibid.,* 7–8. Report to Welles, April 21, 1868.

[47] *Ibid.,* 15.

[48] *Army and Navy Journal,* August 15, 1868, 821.

[49] 40 Cong., 2 sess., *House Exec. Doc. No. 339,* 15–17, which includes reports of Commander William D. Whiting to Welles (June 19, 1868) and of Chief Engineer John S. Albert to Welles (June 24, 1868).

[50] Remarks of E. J. Reed in a discussion of T. Brassey's paper, "On Unarmoured Vessels," *Transactions of the Institution of Naval Architects,* Vol. XVII (1876), 20–21.

# Chapter X

## Line against Staff

[1] Bennett, *Steam Navy,* 609.

[2] Letter from Theodore Zeller to Isherwood, February 5, 1862, "Letters Received," The National Archives, Record Group 19. In this letter Zeller congratulated Isherwood on defeating the Sherman bill aimed at introducing a civilian into the position of engineer in chief and expressed his hope that Isherwood would be as successful "in your efforts to create a Bureau."

[3] Isherwood to Welles, January 29, 1875, Gideon Welles Papers (hereafter referred to as Welles Papers), The New York Public Library.

[4] *Ibid.*

[5] 38 Cong., 1 sess., *House Exec. Doc. No. 1,* xviii–xx, Welles's report of December 7, 1863. Although Fox has been generally given credit for writing this section of the Secretary's report, the ideas were Isherwood's, and most authorities attribute the success of these recommendations to him. Bennett, *Steam Navy,* 654–56; *Power,* XLI, 903; U. S. Navy Department, Bureau of Construction and Repair, *History of the Construction Corps of the United States Navy,* 38.

[6] 38 Cong., 2 sess., *House Exec. Doc. No. 1,* 1214–15.

[7] *Ibid.,* xxxvi–xxxvii.

[8] Isherwood to Welles, February 16, 1865, "Letters Sent to the Secretary of the Navy," The National Archives, Record Group 19.

[9] Isherwood's report of October 12, 1865, 39 Cong., 1 sess., *House Exec. Doc. No. 1,* 308–309.

[10] Isherwood's report of November 3, 1866, 39 Cong., 2 sess., *House Exec. Doc. No. 1,* 178–79.

[11] Bennett, *Steam Navy,* 401.

[12] *Ibid.*, 607. The lack of proper accommodations was undoubtedly caused, in large part, by the amount of space taken up by Isherwood's machinery. For once the engineers could sympathize with the complaints of line officers.

[13] This line-officer resentment and consequent degrading treatment of the engineers was not confined to the American Navy. If anything, it was stronger in the British Royal Navy, in which officers were considerably more caste-conscious than their American counterparts. In addition, although the naval "plumbers" in the British Navy were not all uncouth or uncultured, as a whole they were much more so than American naval engineers. Those men who made a career in American naval engineering were often well educated and socially polished, like Isherwood, Theodore Zeller, James King, and Charles Haswell. It was the voluntary engineer who served only for the duration of the Civil War who often merited the epithet of "greasy mechanic."

[14] George Clymer, *The Principles of Naval Staff Rank* . . . , 36, 39.

[15] *Ibid.*, 37–38.

[16] It was not until 1899 that such an amalgamation successfully took place in the American Navy.

[17] Robert G. Albion's manuscript, "Makers of Naval Policy, 1798–1947," 347.

[18] Welles, *Diary*, II, 233, 236.

[19] *Ibid.*, II, 236, 240–41.

[20] Pages 278, 284.

[21] *Army and Navy Journal*, December 28, 1867, 301.

[22] *Ibid.*, January 11, 1868, 330–31.

[23] "Survey" to the New York *Evening Post*, March 13, 1868, as quoted in Clymer, *The Principles of Naval Staff Rank* . . . , 5.

[24] *Army and Navy Journal*, March 14, 1868, 476–77.

[25] Welles, *Diary*, III, 252–53.

[26] 40 Cong., 2 sess., *Congressional Globe*, 2154.

[27] *Diary*, III, 326.

[28] Fox Papers, The New-York Historical Society.

[29] *Ibid.*

[30] Rear Admiral Daniel Ammen, U. S. Navy, *The Old Navy and the New*, 461–62.

[31] Porter to Chief Engineer W. W. W. Wood, June 18, 1866, published in *The New York Times*, January 29, 1880.

[32] Fox Papers, The New-York Historical Society.

[33] Porter to Washburne, February 6, 1868, David Dixon Porter Papers (hereafter referred to as Porter Papers), in the Naval Historical Foundation Collection (hereafter referred to as NHFC), The Library of Congress.

[34] Welles, *Diary*, III, 384.

[35] Porter Papers, NHFC, The Library of Congress.

[36] Porter to John N. Forney, December 16, 1867, Porter Papers, NHFC, The Library of Congress.

[37] Porter Papers, NHFC, The Library of Congress.

[38] *Ibid.*

[39] Porter to J. A. Supper, an editor of the *Army and Navy Journal*, Porter Papers, NHFC, The Library of Congress.

[40] Porter to Supper, December 31, 1867, Porter Papers, NHFC, The Library of Congress.
[41] Welles, *Diary*, III, 247.
[42] *Ibid.*, III, 253, 563–64.
[43] *Ibid.*, III, 283.
[44] *Ibid.*
[45] Porter Papers, NHFC, The Library of Congress.
[46] *Ibid.*
[47] 40 Cong., 2 sess., *House Exec. Doc. No. 1*, 179–80.
[48] *Ibid.*, 180.
[49] In his capacity as bureau Chief, Isherwood was a civil officer, rather than a naval officer. As line officers, such as Joseph Smith, chief of the Bureau of Yards and Docks, held the rank of rear admiral, it was argued that the staff officers who were also bureau chiefs should hold an equivalent rank, rather than holding a subordinate position by being ranked with commodores. Clymer, *The Principles of Naval Staff Rank . . .* , 164.
[50] Porter Papers, NHFC, The Library of Congress.
[51] *Ibid.*
[52] Porter to Wise, December 20, 1867, Henry August Wise Papers, The New-York Historical Society.
[53] Porter Papers, NHFC, The Library of Congress.

## Chapter XI
### Triumph of the Reactionaries

[1] For a thorough discussion of these tool purchases from the point of view of Roach, see Leonard A. Swann, Jr., *John Roach, Maritime Entrepreneur: The Years as Naval Contractor, 1862–1886*, chapter two.
[2] Welles to Isherwood, June 29, 1868, "Letters to Bureaus of the Navy Department," The National Archives, Record Group 45.
[3] Isherwood to Welles, July 7 and 22, 1868, "Letters Sent to the Secretary of the Navy," The National Archives, Record Group 19.
[4] *Ibid.*
[5] Kelley to Welles, July 28, 1868, quoted in the *Army and Navy Journal*, October 10, 1868, 115.
[6] *Ibid.*, October 3, 1868, 105.
[7] As the other two members of the investigating subcommittee, Aaron F. Stevens and Thomas W. Ferry, rarely attended meetings, Kelley had a free hand in directing the conduct of the investigation.
[8] 40 Cong., 3 sess., *House Report No. 34*, 62–63; Kelley to Welles, July 28, 1868, printed in the *Army and Navy Journal*, October 10, 1868, 115.
[9] 40 Cong., 3 sess., *House Report No. 34*, 19–27.
[10] Roach to Chandler, December 5, 1868, William E. Chandler Papers, The Library of Congress.
[11] 40 Cong., 3 sess., *House Report No. 34*, 28, 59, 62.
[12] *Ibid.*, 76.
[13] Isherwood to Fox, January 7, 1869, Fox Papers, The New-York Historical Society.
[14] 40 Cong., 3 sess., *House Report No. 34*, 27.

[15] Remarks of Representative Samuel J. Randall of Pennsylvania, 40 Cong., 3 sess., *Congressional Globe*, 487.

[16] On June 5, 1866, Zeller wrote to Isherwood, vehemently protesting the rumor that he was a "radical." Maintaining that he had always spoken to Isherwood "with the utmost freedom on political subjects," Zeller reminded his Chief that he was still a Moderate, in favor of a mild reconstruction policy and immediate admission of the seceded states. He totally approved of President Johnson's course, Zeller insisted, and he wanted to dispel any doubts Isherwood might have that he was in favor of the Union "—the Whole Union—A Union in fact & not in name, . . ." and that he was opposed to Radicals such as Stevens and Sumner. Zeller to Isherwood, June 5, 1866, Welles Papers, The Library of Congress.

[17] 40 Cong., 3 sess., *House Exec. Doc. No. 1*, 127. For a description of Isherwood's apparatus, see G. W. Baird, "The Ventilation of Ships, *United States Naval Institute Proceedings*, Vol. VI, 256.

[18] Newton to Ericsson, January 1, 1869, Ericsson Papers, The New-York Historical Society.

[19] Isherwood to Fox, March 5, 1868, Fox Papers, The New-York Historical Society.

[20] Newton to Ericsson, January 6, 1869, Ericsson Papers, The New-York Historical Society.

[21] Ericsson to Welles, January 12, 1869, Ericsson Papers, The New-York Historical Society.

[22] Isherwood to Welles, January 18, 1869, "Letters Sent to the Secretary of the Navy," The National Archives, Record Group 19.

[23] *Ibid.*

[24] Page 296.

[25] Fox Papers, The New-York Historical Society.

[26] *Ibid.*

[27] 40 Cong., 3 sess., *Congressional Globe*, 184.

[28] *Ibid.*, 185. There were 101 abstentions, however.

[29] Porter to Commodore N. L. Case, January 11, 1869, Porter Papers, NHFC, The Library of Congress.

[30] *Diary*, III, 559.

[31] *Ibid.*, III, 441. Porter and Grant had co-operated closely in the siege of Vicksburg, and from that time, the General held the Admiral in great respect. If Grant came to power, Porter knew that he could count on his support and that Grant would follow his recommendations concerning the operations of the Navy Department.

[32] Fox Papers, The New-York Historical Society.

[33] *Diary*, III, 552.

[34] In a letter to Chandler, February 11, 1870, John Roach maintained that he had "positive proof" that King was involved in the tool investigations. William E. Chandler Papers, The Library of Congress.

[35] King to Fox, February 7, 1869, Fox Papers, The New-York Historical Society.

[36] "Letters to Bureaus of the Navy Department," The National Archives, Record Group 45.

[37] *Diary*, III, 549–50, 560.

[38] On July 13, 1866, Isherwood had been confirmed as bureau Chief, his

appointment to take effect for a four-year term ending July 23, 1870. "Confirmations of Appointments of Officers," The National Archives, Record Group 24.

[39] Isherwood to Fox, March 9, 1869, Fox Papers, The New-York Historical Society.

[40] Borie to Commandants, March 13, 1869, "Letters to Officers," The National Archives, Record Group 45.

[41] *Senate Executive Journal,* Vol. XVII, 5, 6, 8.

[42] *Diary,* III, 551-52.

[43] *Ibid.*

## Chapter XII
### *Always the Engineer*

[1] 41 Cong., 2 sess., *House Exec. Doc. No. 1,* Part 1, 142-210.

[2] As reported in a letter from Lenthall to Welles, January 7, 1870, Gideon Welles Papers, The New York Public Library.

[3] "Report on Staff Officers, Mare Island," for December 31, 1869, March 31, 1870, and September 30, 1870; Commodore Parrott's October 3, 1871 letter to Robeson, "Letters from Commandants of Navy Yards and Shore Stations: Mare Island Navy Yard, 1869-71," The National Archives, Record Group 45.

[4] "Officers' Letters," The National Archives, Record Group 45.

[5] "Testimony Taken by the Committee on Naval Affairs . . . in Reference to the Administration of the Navy Department," 45 Cong., 3 sess., *House Misc. Doc. No. 21,* 169-72; "Investigation of the Navy Department," 45 Cong., 3 sess., *House Report No. 112,* 42-54.

[6] 47 Cong., 1 sess., *Congressional Record,* 5697.

[7] "Report from the Naval Advisory Board (November, 1881)," The National Archives, Record Group 45; 47 Cong., 1 sess., *House Exec. Doc. No. 1,* Part 3, Appendix.

[8] *The New York Times,* February 23, 1884; 48 Cong., 1 sess., *Congressional Record,* 1387-88, 5853-54, and 6143-44.

[9] Quoted by Representative William H. Calkins, in Congress, July 5, 1884, 48 Cong., 1 sess., *Congressional Record,* 6144.

[10] Porter to Secretary of the Navy Nathan Goff, January 14, 1881, Porter Papers, NHFC, The Library of Congress.

[11] McKay to Whitney, July 29, 1885, William C. Whitney Papers, The Library of Congress.

[12] Interview with Mrs. Madeleine Kerwin.

[13] Rear Admiral John R. Edwards, U. S. Navy, "The American Society of Naval Engineers—Its Origin, Scope and Purpose," ASNE *Journal,* Vol. XXVI (August, 1914), 685-705.

[14] ASNE *Journal,* Vol. X, 440-73.

[15] As quoted by Clark Fisher, ASNE *Journal,* X, 466.

# Bibliography

There is no collection of Isherwood papers in existence. His official life is thoroughly documented in Navy Department records in the United States National Archives, in the manuscript collections of his contemporaries, and in his own voluminous writings, which include over 150 articles published between 1850 and 1914. However, there are few remaining letters or other manuscript records which reveal anything of his personal life. All of his own published work, of which only a small portion is listed in this bibliography, was of a professional and largely scientific nature.

Robert G. Albion's *Naval and Maritime History: An Annotated Bibliography*, third edition, revised and expanded (Mystic, Connecticut, 1963), is a comprehensive and well-organized selection of unpublished scholarly manuscripts, public documents, published monographs, and other valuable secondary works. Brief critical evaluations of many items provide the student in naval history with a sound introduction to the field. Charles T. Harbeck's compilation, *A Contribution to the Bibliography of the History of the United States Navy* (Cambridge, Massachusetts, 1906), though quite selective in the choice of items, includes descriptive comments on many of them. Robert W. Neeser's *Statistical and Chronological History of the United States Navy, 1775–1907*, 2 vols. (New York, 1909), is not critical but has great value in its exhaustive listing in volume one of manuscripts and published materials.

The Gardner Weld Allen Collection of Naval History in the Harvard College Library, Harvard University, is rarely used, but is a boon to the scholar in search of otherwise inaccessible material on American naval history. It contains a large, comprehensive, and thoroughly indexed assortment of articles and pamphlets relating to all phases of the United States Navy, especially for the nineteenth-century period.

## PRIMARY SOURCES

### I. Manuscripts

A. United States National Archives

Utilizing material on naval history in the National Archives has been greatly simplified in recent years by the issue of preliminary inventory lists for certain major groups of papers. Three of these lists which have provided immeasurable aid in the search for material on Isherwood's official career are: James R. Masterson, compiler, "Preliminary Checklist of the Naval Records Collection of the Office of Naval Records and Library, 1775–1910," an unpublished list compiled at the National Archives in 1945;

Virgil E. Baugh, *Preliminary Inventory of the Records of the Bureau of Naval Personnel,* Washington, D. C., 1960; and Elizabeth Bethel, *et al,* compiler, *Preliminary Inventory of the Records of the Bureau of Ships,* Washington, D. C., 1961.

The files of the Navy Department, especially those of the Bureau of Steam Engineering, are incomplete, as a result of several fires during the latter half of the nineteenth century which destroyed a large portion of the official records. Those which remain have been deposited and catalogued in The National Archives without any attempt to sort or rearrange material. With these limitations, the preliminary inventory lists present a detailed description of Navy Department records. This description includes the contents of each category, the time period covered, and the amount of material, indicated both by the number of bound volumes and by the amount of shelf space occupied.

The Naval Records Collection of the Office of Naval Records and Library contains most of the bound records that originated in the office of the Secretary of the Navy prior to 1886, and for naval historians this is the most valuable group of documents in The National Archives.

The preliminary inventories place archival material into Record Groups which are permanent, and then subdivide them into file groups which are still tentative. In citing this material, I have followed the form suggested by The National Archives and have included only the Record Group number in all but a few cases. The following list of items, however, includes the current file numbers as they are listed and described in the preliminary inventories.

1. *Records of the Bureau of Ships* (Record Group 19).

   *File No.*
   - 61 Records of the Bureau of Construction and Repair, "Letters Received from the Secretary of the Navy" (January, 1861–December, 1869).
   - 963 Records of the Bureau of Engineering, "Letters Sent to the Secretary of the Navy" (1862–69).
   - 965 Records of the Bureau of Engineering, "Letters Sent to Engineers" (July, 1862–69).
   - 970 Records of the Bureau of Engineering, "Letters Received" (July, 1861–July, 1868).
   - 1072 "Steam Logs of Naval Vessels" (USS *Wampanoag*).

2. *Records of the Bureau of Naval Personnel* (Record Group 24).

   *File No.*
   - 38 "Letters Sent to Officers of the Engineer Corps" (1850–69).
   - 118 "Logs of United States Naval Ships and Stations" (Logbook of the USS *Wampanoag*).
   - 148 "Testimonial Letters Concerning Engineers" (1838–69).
   - 152 "Reports of Examining Boards, Engineer Corps, USN" (1849–69).
   - 159 "Confirmations of Appointments of Officers" (1843–69).

182 "Registers of Officers of the Engineer Corps" (1859–69).
193 "Abstracts of Service Records of Naval Officers."
201 "Reports of Line and Staff Officers" (1864–70).
275 "Letters Received from Lieutenant Commanders and Other Officers" (1865–69).
277 "Registers of Letters Received by the Office of Detail" (1865–69).

3. *Naval Records Collection of the Office of Naval Records and Library* (Record Group 45).

*File No.*
1 "Letters to Officers" (1842–69).
5 "Letters to Congress" (1861–69).
13 "Letters to Bureaus of the Navy Department" (1842–69).
17 "Letters to Naval Academy, Commanding and Other Officers" (November, 1869–August, 1884).
22 "Letters from Officers of Rank below That of Commander" (1842–69).
30 "Letters from Officers Commanding Squadrons-South Atlantic Blockading Squadron" (1862–63).
32 "Letters from Bureaus of the Navy Department" (1842–69).
34 "Letters from Commandants of Navy Yards and Shore Stations" (1848–69).
115 "Register of Engineer Officers" (1842–61).
120 "Biographies of Naval Officers" (1865).
188 "Letters from Engineers-in-Chief" (1847–50).
199 "Report from the Naval Advisory Board" (November, 1881).
286 "Register of Engineer Officers" (1843–69).
287 "List of Vessels Serving in the United States Navy during the Civil War."
294 "Records of Proceedings of General Courts Martial and Courts of Inquiry" (No. 3151, G. B. N. Tower).
392 "Logs, Journals, and Diaries of Officers of the United States Navy at Sea" (No. 87, "Medical Journal of Assistant Surgeon R. P. Daniel aboard the USS *San Jacinto,* on a Cruise from New York to Hongkong and back to New York, October 25, 1855–August, 1858," typed copy prepared by the State Office, Jacksonville, Florida, Historical Records Survey, W.P.A., 1938).
464 "Subject File, 1775–1910: Group A, Naval Ships, Design, Construction, etc.; Group E, Engineering."

B. Collections of Private Papers

William E. Chandler Papers, The Library of Congress.
Material on Philadelphia tool investigations and on the Porter-Borie regime in the Navy Department, in letters from John Roach to Chandler.

William Conant Church Papers, The New York Public Library.
   *Army and Navy Journal* correspondence.
William C. Church Papers, The Library of Congress.
Henry Francis du Pont of Winterthur Collection of Manuscripts, Eleutherian Mills Historical Library, Greenville, Wilmington, Delaware.
   Number of official letters from Isherwood to Rear Admiral S. F. du Pont.
John Ericsson Papers, The New-York Historical Society.
   Large collection, mainly his business files. Many letters to and from Stimers, Newton, and Griswold. Official correspondence with Isherwood's bureau.
John Ericsson Papers, The Library of Congress.
Gustavus Vasa Fox Papers, The New-York Historical Society.
   Undoubtedly the most important single collection in Civil War naval history. Mostly letters to Fox, many of which are "private and confidential." Extensive correspondence with Stimers, Ericsson, Porter, Gregory, Eads, Lenthall, Welles, J. W. King, Grimes, Du Pont, and Isherwood, among others.
Louis M. Goldsborough Papers, The Library of Congress.
Louis M. Goldsborough Papers, The New York Public Library.
John T. Hawkins, Second Assistant Engineer, U. S. Navy, Memorandum Book, 1863–64, The New York Public Library.
Benjamin F. Isherwood file, American Society of Mechanical Engineers.
Abraham Lincoln Papers (R. T. Lincoln Collection), The Library of Congress.
R. W. Meade, II, Papers, The New-York Historical Society.
David Dixon Porter Papers, Naval Historical Foundation Collection, The Library of Congress.
   Excellent for line-staff controversy.
David Dixon Porter Papers, The Library of Congress.
   Separate collection from that mentioned above. Some material on line-staff controversy.
David Dixon Porter Papers, The New-York Historical Society.
Robert Henry Thurston Papers, Collection of Regional History and University Archives, Cornell University.
   Several letters from Isherwood in the 1880's.
Aaron Vail Collection, The New-York Historical Society.
Gideon Welles Papers, The New York Public Library.
   Small collection but very good for period after 1868. Letters from Lenthall on changes in the Navy Department under Porter.
Gideon Welles Papers, The Library of Congress.
   Extensive collection. Good for criticism of Navy Department under Porter.
William C. Whitney Papers, The Library of Congress.
Henry August Wise Papers, The New-York Historical Society.

## C. Unpublished Studies

Albion, Robert G. "Makers of Naval Policy, 1798–1947," Washington, D. C., 1950.
  A two-volumed typescript study of American naval administration prepared in the Office of Naval History. Microfilm copies available at the Harvard University Library and elsewhere. The first volume covers the period up to World War II. See chapters seven through nine for congressional relations with the Navy and for the formulation of internal policy.

Haugen, Rolf Nordahl Brun. "The Setting of Internal Administrative Communication in the United States Naval Establishments, 1775–1920," Cambridge, Harvard University, 1953.
  Typescript doctoral dissertation.

## II. Published Material

### A. Public Documents

*Congressional Globe, Containing the Debates and Proceedings,* . . . (known as *Congressional Record, Containing the Proceedings and Debates,* . . . after December 1, 1873). Washington, D. C.

*Official Records of the Union and Confederate Navies in the War of the Rebellion.* Ser. 1, 27 vols.; Ser. 2, 3 vols.; and General Index. Washington, D. C., 1895–1927.

*Register of the Commissioned, Warrant, and Volunteer Officers of the Navy of the United States* . . . , Washington, D. C., 1840–70.

U. S. Congress. *Journal of the Executive Proceedings of the Senate of the United States of America.* Vols. XI–XIV. Washington, D. C.

U. S. Congressional Documents (listed chronologically)

  31 Cong., 1 sess., "Joshua Follansbee and B. F. Isherwood" (to accompany H. R. bill No. 16) *House Report No. 13.*

  32 Cong., 1 sess., "Memorial of Engineers of the Navy Praying a Reorganization of the Corps to Which They Belong," *Sen. Misc. Doc. No. 45.*

  32 Cong., 1 sess., "Report of the Secretary of the Navy—Appendix D, Letter of Engineer Isherwood . . . Relating to Improvements in the Boilers of the Steam Frigate Mississippi," *House Exec. Doc. No. 2.*

  32 Cong., 2 sess., "Navy Steamers," *House Exec. Doc. No. 63.*

  33 Cong., 1 sess., "Steam Navy of the United States," *House Exec. Doc. No. 65.*

  36 Cong., 2 sess., "Report of the Secretary of the Navy," *Sen. Exec. Doc. No. 1.*
    Includes report of the board to examine conversion of sailing warships to steamers.

  36 Cong., 2 sess., "The Naval Establishment," *Sen. Exec. Doc. No. 4.*

  37 Cong., 1 sess., "Report of the Secretary of the Navy," *Sen. Exec. Doc. No. 1.*
    Welles's first report, dated July 4, 1861.

37 Cong., 1 sess., "Destruction of the United States Vessels and Other Property at Norfolk, Virginia," *House Exec. Doc. No. 11.*

37 Cong., 2 sess., "Report of the Secretary of the Navy," *Sen. Exec. Doc. No. 1.*
   Welles's report of December 2, 1861.

37 Cong., 2 sess., "Document Establishing Two Bureaus in the Navy Department" (to accompany S. bill No. 171), *Sen. Misc. Doc. No. 46.*

37 Cong., 2 sess., "Surrender and Destruction of Navy Yards, etc.," *Sen. Report. No. 37.*
   Highly critical report of the abandonment of the Norfolk Navy Yard.

37 Cong., 2 sess., "Iron-Clad Ships, Ordnance . . . ," *House Misc. Doc. No. 82.*
   Letters from Welles to House Committee on Naval Affairs, March 25, 1862 and June 9, 1862, requesting government manufacturing facilities.

37 Cong., 3 sess., "Letter of the Secretary of the Navy Communicating . . . Information in Relation to the War Steamers *Ossipee* and *Pensacola*," *Sen. Exec. Doc. No. 45.*
   February 20, 1863 letter in answer to Senate resolution of January 15, 1863, initiated by Edward Dickerson.

37 Cong., 3 sess., "Report of the Secretary of the Navy," *House Exec. Doc. No. 1.*
   Welles's report of December 1, 1862.

38 Cong., 1 sess., "Select Committee on Naval Supplies," *Sen. Report No. 99.*
   Report and testimony presented to the committee, February to June, 1864. Testimony by Isherwood and Lenthall on their bureaus' contract procedures.

38 Cong., 1 sess., "Report of the Secretary of the Navy," *House Exec. Doc. No. 1.*
   Welles's report of December 7, 1863.

38 Cong., 2 sess., "Light-Draught Monitors," *Report of the Joint Committee on the Conduct of the War, Pt. 3, Sen. Report No. 142.*

38 Cong., 2 sess., "Report of the Secretary of the Navy," *House Exec. Doc. No. 1.*
   Welles's report of December 5, 1864. Includes Isherwood's detailed bureau report of November 28, 1864, on training engineers at the Naval Academy, on all vessels built with his machinery, on the *Pensacola* failure, and on fuel-oil experiments.

38 Cong., 2 sess., "Marine Engines," *House Report No. 8.*
   Report of January 30, 1865, supporting Isherwood's machinery designs and his operation of the Bureau of Steam Engineering.

39 Cong., 1 sess., "Report of the Secretary of the Navy," *House Exec. Doc. No. 1.*
   Welles's report of December 4, 1865. Includes Isherwood's bureau report of October 12, 1865, which argues for government manufacturing facilities and discusses steam-expansion experiments.

# Bibliography 273

39 Cong., 2 sess., "Report of the Secretary of the Navy," *House Exec. Doc. No. 1.*
    Welles's report of December 3, 1866. Includes Isherwood's bureau report of November 3, 1866, which discusses the effect of postwar retrenchment.

40 Cong., 2 sess., "Report of the Secretary of the Navy," *House Exec. Doc. No. 1.*
    Welles's report of December 2, 1867. Includes Isherwood's highly controversial report of October 25, 1867, asking for higher rank for engineers.

40 Cong., 2 sess., "Letter from the Secretary of the Navy . . . Relative to the Trial of the United States Steamer *Wampanoag*, and Vessels of That Class . . . ," *House Exec. Doc. No. 339.*
    July 18, 1868 letter which gives trial data of the *Wampanoag*, the *Ammonoosuc*, the *Madawaska*, the *Chattanooga*, and the *Idaho*.

40 Cong., 3 sess., "Report of the Secretary of the Navy," *House Exec. Doc. No. 1.*
    Welles's report of December 7, 1868. Includes Isherwood's bureau report of October 22, 1868, which gives an excellent comparison of the *Wampanoag* with competing vessels.

40 Cong., 3 sess., "Purchase of Tools by Theodore Zeller," *House Report No. 4.*
    Minority report of Kelley subcommittee on January 5, 1869.

40 Cong., 3 sess., "Purchase of Tools by Theodore Zeller," *House Report No. 34.*
    Report by Committee on Naval Affairs, February 26, 1869, exonerating Zeller.

41 Cong., 2 sess., "Report of the Secretary of the Navy," *House Exec. Doc. No. 1.*

41 Cong., 2 sess., "Causes of the Reduction of American Tonnage," *House Report No. 29.*

41 Cong., 2 sess., "Change of Name of United States Vessels," *House Exec. Doc. No. 92.*

41 Cong., 3 sess., "Report of the Secretary of the Navy," *House Exec. Doc. No. 1, Pt. 3.*

42 Cong., 2 sess., "Charges against the Navy Department," *House Misc. Doc. No. 201.*
    Includes material on Isherwood's administration of the Bureau of Steam Engineering, 1863–68.

42 Cong., 2 sess., "Experiments on the Coals of the Pacific Coast," *House Exec. Doc. No. 206.*

42 Cong., 3 sess., "Report of the Secretary of the Navy," *House Exec. Doc. No. 1, Pt. 3.*

43 Cong., 2 sess., "Report of the Secretary of the Navy," *House Exec. Doc. No. 1, Pt. 3.*

44 Cong., 1 sess., "Investigation by the Committee on Naval Affairs, . . . ," *House Misc. Doc. No. 170, Pt. 9.*
    Includes information on Isherwood and the Martin boiler.

44 Cong., 1 sess., "Investigations of the Navy Department," *House Report No. 784.*

44 Cong., 1 sess., "Report of the Secretary of the Navy," *House Exec. Doc. No. 1, Pt. 3.*

45 Cong., 3 sess., "Testimony Taken by the Committee on Naval Affairs ... in Reference to the Administration of the Navy Department," *House Misc. Doc. No. 21.*

45 Cong., 3 sess., "Investigation of the Navy Department," *House Report No. 112.*

47 Cong., 1 sess., "Report of the Secretary of the Navy," *House Exec. Doc. No. 1, Pt. 3.*

47 Cong., 1 sess., "Condition of the Navy," *House Exec. Doc. No. 30, Pts. 1, 2, 3, and 4.*

47 Cong., 1 sess., "Construction of Vessels of War for the Navy," *House Report No. 653.*

47 Cong., 2 sess., "List of Vessels Stricken from the Navy Register," *House Exec. Doc. No. 66.*

49 Cong., 2 sess., "Report of the Secretary of the Navy," *House Exec. Doc. No. 1, Pt. 3.*

U. S. Navy Department Documents

*David W. Taylor Model Basin, The,* Washington, D. C., 1939.

Bureau of Construction and Repair. *History of the Construction Corps of the United States Navy.* Washington, D. C., 1937.

Office of the Chief of Naval Operations, Naval History Division, *Dictionary of American Naval Fighting Ships.* Washington, D. C., 1959–.

"Report Made to the Bureau of Steam Engineering, Navy Department, August 9, 1882, by B. F. Isherwood ... on the Vedette Boats Constructed for the British and French Navies by the Herreshoff Manufacturing Company, at Bristol, R.I.," Washington, D. C., 1882.

"Report Made to the Bureau of Steam Engineering, Navy Department, March 3, 1883, by B. F. Isherwood ... on the Hull, Engine, and Boiler of the Steam-Yacht 'Siesta,' Constructed by the Herreshoff Manufacturing Company, ... ," Washington, D. C., 1883.

"Report of a Board of United States Naval Engineers on the Mallory Steering and Propelling Screw, as Applied to the United States Torpedo Boat *Alarm,* and on the Experiments with It in That Vessel, Made to the Bureau of Steam Engineering, Navy Department, January 31, 1882," Washington, D. C., 1882.

"Report of a Board of United States Naval Engineers on the Herreshoff System of Motive Machinery as Applied to the Steam-Yacht *Leila,* and on the Performance of That Vessel, Made to the Bureau of Steam Engineering, Navy Department, June 3, 1881," Washington, D. C., 1881.

"Report of Chief Engineer Isherwood to the Secretary of the Navy on the Zero-Motor, New York, March 19, 1881," Washington, D. C., 1881.

Thompson, M. S., comp. *General Orders and Circulars Issued by the Navy Department from 1863 to 1887....* Washington, D. C., 1887.

B. Books and Articles by Isherwood

"Association Notes—Letter from Chief Engineer B. F. Isherwood, USN," *Journal of the American Society of Naval Engineers,* Vol. XXVI (May, 1914), 678–80.

*Description and Illustration of Spaulding and Isherwood's Plan of Cast Iron Rail and Superstructure for Rail-roads.* New York, 1842.

*Engineering Precedents for Steam Machinery; Embracing the Performance of Steamships, Experiments with Propelling Instruments, Condensers, Boilers,* ... Vols. I and II. New York, 1859.

"Expansion of Steam in the Steam Engine," *Journal of the Franklin Institute,* Vol. CVI (July, 1878), 1–12.

*Experimental Researches in Steam-Engineering,* Vol. I, Philadelphia, 1863; Vol. II, Philadelphia, 1865.

"Experiments Made at the Mare Island Navy-Yard, California, with Different Screws Applied to the United States Steam Launch No. 4, to Ascertain Their Relative Propelling Efficiency," *Journal of the Franklin Institute,* Vol. XCIX (April, 1875), 278–85; (May, 1875), 342–49; (June, 1875), 393–405; (July, 1875), 34–40; (August, 1875), 104–16; (September, 1875), 165–77; (October, 1875), 253–62; (December, 1875), 393–400.

"Investigation of the Comparative Merits of the Perpendicular and Radial Paddle Wheels for Sea-going Vessels," *Journal of the Franklin Institute,* Vol. L (August, 1850), 134–39; (September, 1850), 181–88.

"Notes on Indicator Diagrams from the U. S. War Steamer 'Spitfire,'" *Journal of the Franklin Institute,* Vol. LI (February, 1851), 91–95, 168–71.

"Practical Description of the Timber Bridges..." in John Weale, ed., *Ensamples of Railway Making; Which, although not of English Practice, Are Submitted with Practical Illustrations, to the Civil Engineer, and the British and Irish Public.* London, 1843.

"Practical Description of the Timber Bridges, . . . on the Utica and Syracuse Railway, in the United States," Vol. II in John Weale, ed., *The Theory, Practice, and Architecture of Bridges,* . . . London, 1843.

"Remarks on Nystrom's Screw Propeller," *Journal of the Franklin Institute,* Vol. LII (July, 1851), 42–48.

"Sloop-of-War 'Wampanoag': A Once Famous, but Long Forgotten, United States Cruiser, The," *Cassier's Magazine,* Vol. XVIII (August, 1900), 282–89; (September, 1900), 403–24.

"Trial Trip of the U. S. Screw Propeller Steamship of War, 'San Jacinto,'" *Journal of the Franklin Institute,* Vol. LII (November, 1851), 339–51.

"United States Steamer 'Water Witch,' The," *Journal of the Franklin Institute,* Vol. LV (March, 1853), 178–88.

"U. S. Screw Steamship 'Princeton,'" *Journal of the Franklin Institute,* Vol. LV (June, 1853), 377–85; Vol. LVI (July, 1853), 43–51.

"Water-Tube and Fire-Tube Boilers," *Journal of the Franklin Institute,* Vol. CVII (January, 1879), 14–24.

C. Diaries, Memoirs, and Letters

Ammen, Daniel, Rear Admiral, U. S. Navy. *The Old Navy and the New.* Philadelphia, 1891.

Belknap, George E., Captain, U. S. Navy. "Reminiscent of the 'New Ironsides' off Charleston," *The United Service* (January, 1879), 63–82.

Buell, Augustus C. *The Memoirs of Charles H. Cramp.* Philadelphia, 1906.
  Usually unreliable, either when quoting Cramp directly or in his own narrative.

Davis, Henry Winter. *Speeches and Addresses Delivered in the Congress of the United States,* ... New York, 1867.
  A good sampling of one of the Navy Department's inveterate antagonists during the Civil War.

Drayton, Percival. "Naval Letters from Captain Percival Drayton, 1861–1865," *Bulletin of The New York Public Library,* Vol. X (November, 1906), 587–625; (December, 1906), 639–81.
  Contains some pertinent criticisms of Ericsson's monitor designs.

Fox, Gustavus. *Confidential Correspondence of Gustavus Fox, Assistant Secretary of the Navy, 1861–1865.* Ed. by Robert Means Thompson and Richard Wainwright. 2 vols. New York, 1918–19.
  A small and not necessarily the best portion of the Fox Papers at the New-York Historical Society. Very little material on Isherwood.

Harris, Townsend. *The Complete Journal of Townsend Harris; First American Consul General and Minister to Japan.* Ed. by Mario Emilio Cosenza. Garden City, New York, 1930.

Heusken, Henry. *Japan Journal: 1855–1861.* Translated and ed. by Jeannette C. Van der Corput and Robert A. Wilson. New Brunswick, New Jersey, 1964.

Keeler, William Frederick, Acting Paymaster, U. S. Navy. *Aboard the USS Monitor: 1862.* Ed. by Robert W. Daly. Annapolis, 1964.

Lincoln, Abraham. *The Collected Works of Abraham Lincoln.* Ed. by Roy Basler. Vol. IV (1860–61). New Brunswick, New Jersey, 1953.

Mahan, A. T. *From Sail to Steam: Recollections of Naval Life.* New York, 1907.

Porter, David D., Admiral, U. S. Navy. *The Naval History of the Civil War.* New York, 1886.
  Lengthy and often inaccurate account, mainly of battles and fleet action.

Russell, William Howard. *My Diary North and South.* 2 vols. London, 1863.
  Contemporary and occasionally acrid account by a British journalist on the conduct of the Civil War in its early stages.

Semmes, Raphael, Lieutenant, U. S. Navy. *Service Afloat and Ashore during the Mexican War.* Cincinnati, 1851.

Shock, William H., Chief Engineer, U. S. Navy. "Narrative of a First Cruise," *The United Service* (February–June, 1893), 121–30; 256–70; 331–39; 467–72; 533–39.

Taylor, Reverend Fitch W., U. S. Navy. *The Broad Pennant: Or, A Cruise in the United States Flag Ship of the Gulf Squadron, during the Mexican Difficulties,* ... New York, 1848.

Welles, Gideon. *Diary of Gideon Welles, Secretary of the Navy under Lincoln and Johnson.* Ed. by Howard K. Beale. 3 vols. New York, 1960.

A new edition of the diary which indicates changes from Welles's original draft. Retains the same pagination as the earlier Morse edition. An invaluable source for Civil War political and naval history, full of colorful and occasionally fiery observations.

Wood, William Maxwell, M. D., U. S. Navy. *Fankwei: Or, The* San Jacinto *in the Seas of India, China and Japan.* New York, 1859.

D. Contemporary Pamphlets

*Celebration of the Centennial Anniversary of the Founding of the Albany Academy, May 24, 1913, The.* Albany, 1914.

*Celebration of the Semi-Centennial Anniversary of the Albany Academy, June 23, 1863, The.* Albany, 1863.

*Celebration of the Seventy-Fifth Anniversary of the Founding of the Albany Academy; October 25, 1888, The.* Albany, 1888.

[Clymer, George.] "A Surgeon in the U. S. Navy." *The Principles of Naval Staff Rank; and Its History in the United States Navy, for over Half a Century.* n. p., 1869.

Dickerson, Edward N. *The Navy of the United States: An Exposure of Its Condition, and the Causes of Its Failure.* New York, 1864.

The most thoroughgoing and damning of all the pamphlets attacking Isherwood.

——. *The Steam Navy of the United States: Its Past, Present, and Future.* New York, 1863.

A public letter to Welles warning him of Isherwood's influence on the Navy.

"Engineer, An". *A Brief Sketch of Some of the Blunders in the Engineering Practice of the Bureau of Steam Engineering in the U. S. Navy.* New York, 1868.

Probably written by Isaac Newton.

*Examination of the Report of the Naval Committee . . . on Naval Steam Machinery, Showing the Misrepresentations Resorted to by the Chief of the Bureau of Steam Engineering before That Committee, An.* New York, 1866.

Contains six letters to *The New York Times* and also quotations from several New York and London papers criticizing Isherwood.

*Facts in Relation to the Official Career of B. F. Isherwood, Chief of the Bureau of Steam Engineering of the Navy Department.* Philadelphia, 1866.

A series of essays attacking Isherwood.

Hawley, Gideon. *Address Delivered at the Public Exercises of the Albany Academy, in the Second Dutch Church, August 6, 1835.* Albany, 1835.

McPherson, John R. *Relics of Ex-Secretary Robeson's "Navy," (Speech in the U. S. Senate, March 29, 1880).* Washington, D. C., 1880.

[Newton, Isaac, Chief Engineer, U. S. Navy.] *Some Facts of History, Showing the Official Career of the Engineers-in-Chief of the Navy, and Their Deleterious Influence upon the Navy Itself and the Mechanical Interests of the Country.* New York, 1868.

*Official Reports on the Inefficiency of the U.S.S.* Wampanoag *and Mr.*

*Isherwood's Defence, The: Remarks on Mr. Isherwood's Defence.* New York, 1868.

"Pro Bono Publico." *An Appeal to His Excellency Andrew Johnson, President of the United States; The Hon. Gideon Welles, Secretary of the Navy; and to the Hon. Members of the Senate and House of Representatives, of the United States, in Congress Assembled.* n.p., 1866.
    Strong attack on Isherwood by an "ex-naval engineer," probably Newton.

Shultes, Edwin Matthias, III, ed. *The Celebration of the One Hundred and Twenty-Fifth Anniversary of the Founding of Albany Academy; June 2, 3, 4, 1938.* Albany, 1938.

*Statutes of the Albany Academy, Revised and Passed, October 9, 1829. Reprinted, with Amendments and Alterations, September, 1843.* Albany, 1843.

"W.S.W.R." *A Brief History of an Existing Controversy on the Subject of Assimilated Rank in the Navy of the United States.* Philadelphia, 1850.

E. Newspapers and Periodicals

*Engineer, The* (London), 1861–69.

*Engineering: An Illustrated Weekly Journal* (London), 1866–69.

*Journal of the Franklin Institute* (Philadelphia), 1850–99.

*New York Times, The,* 1861–1920.

*Scientific American,* 1861–69.

*United States Army and Navy Journal and Gazette of the Regular and Volunteer Forces, The,* August 29, 1863–1886.

## SECONDARY SOURCES

### I. Biographical Articles on Isherwood

"A.S.M.E. Presented with Bust of Admiral Isherwood," *Power,* Vol. XLVII (January 15, 1918), 104.

Baird, George W. "Chief Engineer Benjamin F. Isherwood," *Journal of the American Society of Naval Engineers,* Vol. XXVII (August, 1915), 733–36.

"Dedication of a Bronze Bust of Chief Engineer Isherwood, U. S. Navy," *Journal of the American Society of Naval Engineers,* Vol. X (May, 1898), 440–73.
    Includes a number of reminiscences by naval engineers who worked for Isherwood during the Civil War.

Dyson, George W. "Benjamin Franklin Isherwood," *United States Naval Institute Proceedings,* Vol. LXVII (August, 1941), 1138–46.

Edwards, E. J. "The Men Who Made Our Modern Sea Fighters Possible. The Three Persons Most Vitally Responsible for It Still Live—Admirals Isherwood and Melville and Sir William White," *The New York Times Magazine Section* (November 12, 1911), 12.

Kirk, Neville T. "The Father of American Naval Engineering," *United States Naval Institute Proceedings,* Vol. LXXXI (April, 1955), 486–89.

McCormick, Robert R., Colonel. "Benjamin Franklin Isherwood, Hero." Chicago, 1947.

    An address broadcast May 10, 1947, over radio station WGN, Chicago, and the Mutual Broadcasting System.

McFarland, Walter M. "Benjamin F. Isherwood, Engineer and Scientist," *Journal of the American Society of Naval Engineers,* Vol. XXVIII (February, 1916), 1–10.

McKean, F. C. "An Appreciation of a Well-Known Engineer," *Journal of the American Society of Naval Engineers,* Vol. XXVII (November, 1915), 1053–54.

Rankin, Robert H. "Isherwood: Naval Iconoclast; the Fight for the Modern Navy," *Our Navy,* Vol. LVI (March, 1961), 12–13.

"Rear Admiral Isherwood Dead at 92," *Power,* Vol. XLI (June 29, 1915), 903.

Taylor, Frank A. "Benjamin Franklin Isherwood," *Dictionary of American Biography,* Vol. IX, 515–16.

    Used by the U. S. Navy Department as its official biography of Isherwood.

Thurston, R. H. "Benjamin F. Isherwood," *Cassier's Magazine,* Vol. XVIII (August, 1900), 344–52.

    Highly eulogistic but complete account by a friend and former associate of Isherwood.

## II. Books and Articles on Marine Engineering and Naval Architecture

Albert, F. P. "How a Barrel of Olive Oil Saved the Staff of the Navy; An Incident in American Naval History," *Engineering News,* Vol. LVI (December 27, 1906), 682–84.

Alden, John D. "Born Forty Years Too Soon," *American Neptune,* Vol. XXII (October, 1962), 252–63.

    Discussion of the *Roanoake,* converted to a turreted ironclad by Isherwood and Lenthall.

Armytage, W. H. G. *A Social History of Engineering.* London, 1961.

    Inaccurate and superficial on the American Navy in the 1860's.

Baird, G. W. "The Ventilation of Ships," *United States Naval Institute Proceedings,* Vol. VI (1880), 237–63.

Baxter, James Phinney, III. *The Introduction of the Ironclad Warship.* Cambridge, 1933.

    A thorough monograph on the ironclad up to 1862. Material on the American Navy taken from The National Archives and from the Fox, Welles, and Ericsson papers, among others. A model piece of historical research and writing.

Bennett, Frank M. *The Steam Navy of the United States.* Pittsburgh, 1896.

    An invaluable study of American naval engineering in all its aspects up to 1900. Obviously written from official government documents and manuscript records, but there are no citations or a bibliography. As a friend of Isherwood, Bennett is generally sympathetic toward the Engineer in Chief.

———. *The Monitor and the Navy under Steam.* Boston, 1900.

Brassey, T. "On Unarmoured Vessels," *Transactions of the Institution of Naval Architects,* Vol. XVII (1876), 13–28.
See especially the remarks of Sir E. J. Reed on the *Inconstant.*

Brown, Alexander Crosby, Lieutenant, U. S. Naval Reserve. "Monitor-class Warships of the United States Navy," Society of Naval Architects and Marine Engineers *Historical Transactions, 1893–1943* (1945), 330–38.

Cressy, Edward. *A Hundred Years of Mechanical Engineering.* New York, 1937.
A good survey of the evolution of steam engine theory.

Dyson, C. W. "A Fifty Year Retrospect of Naval Marine Engineering," *Journal of the American Society of Naval Engineers,* Vol. XXX (May, 1918), 255–302.
Good general review of American naval engineering from 1863 to 1913.

Edwards, John R., Rear Admiral, U. S. Navy. "The American Society of Naval Engineers—Its Origin, Scope and Purpose," *Journal of the American Society of Naval Engineers,* Vol. XXVI (August, 1914), 685–705.

Emmons, George F., Lieutenant, U. S. Navy, comp. *The Navy of the United States, from the Commencement, 1775 to 1853; with a Brief History of Each Vessel's Service and Fate as Appears upon Record.* Washington, D. C., 1853.

Gurley, Ralph R. "The *Wampanoag,*" *United States Naval Institute Proceedings,* Vol. LXIII (December, 1937), 1732–36.
Draws from Bennett and Isherwood.

Hichborn, Philip, Naval Constructor, U. S. Navy. "Sheathed or Unsheathed Ships?" *United States Naval Institute Proceedings,* Vol. XV (1889), 21–56.

Hovgaard, William. *Modern History of Warships.* London, 1920.
Admirable survey by a leading authority on naval architecture who was the former head of the School of Naval Architecture and Marine Engineering at the Massachusetts Institute of Technology.

Hutton, Frederick Remsen. *A History of the American Society of Mechanical Engineers from 1880 to 1915.* New York, 1915.

King, J. H., and R. S. Cox. "The Development of Marine Watertube Boilers," Society of Naval Architects and Marine Engineers *Historical Transactions, 1893–1943,* 476–516.

King, J. W., Chief Engineer, U. S. Navy. *Report of Chief Engineer J. W. King, United States Navy, on European Ships of War and Their Armament,* . . . 2nd ed. Washington, D. C., 1878.
Critical of Isherwood and the *Wampanoag.* King's *Warships and Navies of the World* (1880) is an enlarged and popularized edition of this official report.

King, W. H., and Chief Engineer J. W. King, U. S. Navy. *Lessons and Practical Notes on Steam, the Steam Engine, Propellers,* . . . 10th ed. New York, 1865.

A popular textbook which defends the use of Mariotte's law against Isherwood's theories of steam expansion.

MacBride, Robert. *Civil War Ironclads: The Dawn of Naval Armor.* Philadelphia, 1962.
General survey based on published works.

Main, Thomas. *The Progress of Marine Engineering: From the Time of Watt until the Present Day.* New York, 1893.
Critical of Isherwood's expansion theories and of his engines. Written by a civilian engineer working for John Roach during the Civil War.

Millis, Walter. "The Iron Sea Elephants," *American Neptune,* Vol. X (January, 1950), 15-32.

Neeser, Robert W. "The Ships of the United States Navy: An Historical Record of Those now in Service and of Their Predecessors of the Same Name, 1776-1915," *United States Naval Institute Proceedings,* Vol. XLI (September-October, 1915), 1623-31.

Neuhaus, Herbert M. "Fifty Years of Naval Engineering in Retrospect," *Journal of the American Society of Naval Engineers,* Vol. L (February, May, August, and November, 1938), 1-38; 240-80; 341-80; 527-64.

Nichols, John F. "The Development of Marine Engineering," Society of Naval Architects and Marine Engineers *Historical Transactions, 1893-1943,* 425-36.

Oliver, John W. *History of American Technology.* New York, 1956.
Brief, laudatory comment on Isherwood's improvements on simple steam-engine design.

Rankine, W. J. Macquorn. "Remarks on Some Experiments Made by Messrs. Isherwood, Stimers, and Long . . . on the Liquefaction of Steam in the Cylinder of an Engine Working Expansively," *Transactions of the Institution of Engineers in Scotland,* Vol. V, 61-69.
Analysis of the results of the Erie experiments.

Reed, E. J. "On Iron-cased Ships of War, . . ." *Transactions of the Institution of Naval Architects,* Vol. IV (1863), 31-50.

Reingold, Nathan. "Science in the Civil War—The Permanent Commission of the Navy Department," *Isis,* Vol. XLIX (September, 1958), 307-18.
Discusses Isherwood and the department-sponsored experiments on steam expansion.

Rossell, H. E. "Types of Naval Ships," Society of Naval Architects and Marine Engineers *Historical Transactions, 1893-1943,* 248-328.

Seaton, A. E. *The Screw Propeller: and other Competing Instruments for Marine Propulsion.* Chapter 18. London, 1909.

Shock, William S. *Steam Boilers: Their Design, Construction, and Management.* New York, 1880.
Written by a chief of the Bureau of Steam Engineering in the American Navy. Gives credit to Isherwood's prior work and to his suggestions for this book.

Singer, Charles, *et al.,* eds. *The Late Nineteenth Century, c. 1850 to c. 1900.* Vol. V in *A History of Technology.* Oxford, 1958.
Attributes the term "atomize" to Isherwood.

Smith, Edgar C. *A Short History of Naval and Marine Engineering.* Cambridge, England, 1938.
> Excellent survey from the English point of view, by a British naval engineer.

Stuart, Charles B., Engineer in Chief, U. S. Navy. *The Naval and Mail Steamers of the United States.* New York, 1853.

Swan, James. "The 'Wampanoag,'" *Transactions of the Society of Naval Architects and Marine Engineers,* Vol. XXXV (1927), 43-54.
> Well-illustrated, thorough discussion of Isherwood's cruiser. Next to Isherwood's own article, the best discussion on the vessel.

Thurston, Robert H. *The Development of the Philosophy of the Steam Engine: An Historical Sketch.* New York, 1889.
> Good discussion of Isherwood's place in the development of steam technology by the leading American authority on steam engineering in the nineteenth century.

———. *A History of the Growth of the Steam-Engine.* New York, 1878.
> No mention of Isherwood, but a valuable contemporary view of the developments in steam theory and practice.

———. *A Manual of the Steam Engine. Part I, Structure and Theory.* 6th revised edition. New York, 1903.
> Brief mention of Isherwood's role in the development of thermodynamic theory.

Tyler, David Budlong. *Steam Conquers the Atlantic.* New York, 1939.

"The U.S.S. Armored Frigate New Ironsides," *Journal of the Franklin Institute,* Vol. LXXXIII (February, 1867), 73-81.

Very, Edward W., Lieutenant, U. S. Navy. *Navies of the World.* New York, 1880.

Ward, J. H., Commander, U. S. Navy. *Steam for the Million—A Popular Treatise on Steam and Its Application to the Useful Arts, Especially to Navigation, . . .* New York, 1860.
> A representative contemporary account by a commander in the U. S. Navy.

Watts, Sir Philip. "Warship Building (1860-1910)," *Transactions of the Institution of Naval Architects,* Vol. LIII, Pt. 2 (1911), 291-337.
> Brief discussion of *Wampanoag-Inconstant* development.

## III. Biographical Works

Barnes, Gilbert H. "Charles B. Stuart," *Dictionary of American Biography,* Vol. XVIII, 163.

Bigelow, Donald Nevius. *William Conant Church and the Army and Navy Journal.* New York, 1952.
> Sympathetic to Isherwood, shows Church-Ericsson relationship and reasons for attacks on Isherwood.

Callahan, Edward W., ed. *List of Officers of the Navy of the United States and of the Marine Corps from 1775 to 1900.* New York, 1901.
> Service records taken from the official records of the Navy Department by the Registrar of the Bureau of Navigation.

Church, William Conant. *The Life of John Ericsson.* 2 vols. New York, 1890.
    Strong bias toward Ericsson, but valuable in its detail and for frequent quotations from Ericsson's letters. Still the best biography on Ericsson.

Coulson, Thomas. *Joseph Henry: His Life and Work.* Princeton, 1950.

Davis, Charles H. *Life of Charles Henry Davis, Rear Admiral, 1807–1877.* Boston, 1899.
    Unsatisfactory biography by his son. Little mention of his work in the Navy Department during the Civil War.

Durand, William F. "John Ericsson—Navies of War and Commerce," *Beacon Lights of History* (ed. by John Lord), Vol XIV. New York, 1902.
    A perceptive delineation of Ericsson's character.

———. *Robert Henry Thurston, A Biography.* New York, 1929.
    Briefly mentions Thurston-Isherwood relationship.

Gilman, William H. *Melville's Early Life and REDBURN.* New York, 1951.

Griffis, William Elliott. *Matthew Calbraith Perry: A Typical American Naval Officer.* Boston, 1887.

Hamersly, Lewis Randolph, comp. *The Records of Living Officers of the U. S. Navy and Marine Corps.* 5th ed. Philadelphia, 1894.

Jahns, Patricia. *Matthew Fontaine Maury and Joseph Henry: Scientists of the Civil War.* New York, 1961.

Jones, Charles C., Jr. *The Life and Services of Commodore Josiah Tatnall.* Savannah, Georgia, 1878.

Knott, H. W. Howard. "Edward Nicoll Dickerson," *Dictionary of American Biography,* Vol. V, 288–89.

McFarland, Walter M. "Honored Names in Marine Engineering," *Journal of the American Society of Naval Engineers,* Vol. XXXIX (May, 1927), 207–26.

Morris, Maud Burr. "The Lenthall Houses and Their Owners," *Records of the Columbia Historical Society, Washington, D. C.* Vols. XXXI and XXXII, 1–35.
    A brief description of John Lenthall's private life. No mention of Isherwood.

Norris, Walter B. "John Lenthall," *Dictionary of American Biography,* Vol. XI, 173.
    No mention of Isherwood; good only for Lenthall's official career as a naval constructor.

Richardson, Leon Burr. *William E. Chandler—Republican.* New York, 1940.

Salter, William. *The Life of James W. Grimes.* New York, 1876.
    Sketchy. Mostly a stringing together of Grimes's letters and public speeches.

Soley, James Russell. *Admiral Porter.* New York, 1903.

Swann, Leonard A., Jr. *John Roach, Maritime Entrepreneur: The Years as Naval Contractor, 1862–1886.* Annapolis, 1965.

"Theodore Zeller," *Journal of the American Society of Naval Engineers,* Vol. XIII (1901), 819.

Unsigned obituary, in all probability written by Isherwood.

West, Richard S., Jr. *Gideon Welles: Lincoln's Navy Department.* New York, 1943.
> Based on the various collections of Welles papers. Fairly good for general description of naval administration during and after the Civil War. Little on Isherwood.

——. *The Second Admiral: A Life of David Dixon Porter, 1813–1891.* New York, 1937.
> Good only through the Civil War period. Inadequate for the remainder of Porter's career, although based on the various collections of Porter papers. Does not recognize the actual extent of political maneuvering by Porter in the post-Civil War years.

White, Ruth. *Yankee from Sweden: The Dream and the Reality in the Days of John Ericsson.* New York, 1960.
> A brief, popularized account of the inventor's career, written from the Ericsson manuscript collections. More dispassionate than Church's book, but not as thorough or as informative.

## IV. Works on Naval and Maritime History

Albion, Robert G. "The Naval Affairs Committees, 1816–1947," *United States Naval Institute Proceedings,* Vol. LXXVIII (November, 1952), 1227–37.
> An abbreviated version of the account presented in the author's "Makers of Naval Policy, 1798–1947."

Anderson, Bern. *By Sea and by River: The Naval History of the Civil War.* New York, 1962.
> A general view of naval strategy for the period, based on published government documents.

Beers, Henry P. "The Development of the Office of the Chief of Naval Operations, Part I," *Military Affairs,* Vol. X (Spring, 1946), 40–68.

Boynton, Charles B. *The History of the Navy during the Rebellion.* 2 vols. New York, 1867.
> A strong Anglophobic account, written by a chaplain and utilizing Navy Department records.

Brassey, Sir Thomas. *The British Navy: Its Strength, Resources, and Administration.* 5 vols. London, 1882–83.

Bruce, Robert V. *Lincoln and the Tools of War.* Indianapolis, 1956.

Bryant, Samuel W. *The Sea and the States—A Maritime History of the American People.* New York, 1947.
> Enthusiastic account based on a minimum of research. Good for Isherwood-Porter relationship.

Conner, Philip Syng Physick. *The Home Squadron under Commodore Conner in the War with Mexico, Being a Synopsis of Its Services.* Philadelphia, 1896.

Dalzell, George W. *The Flight from the Flag: The Continuing Effect of the Civil War upon the American Carrying Trade.* Chapel Hill, North Carolina, 1940.

Sees the *Wampanoag* as an influence on settlement of the Alabama claims.

Davis, George T. *A Navy Second to None: The Development of Modern American Naval Policy.* New York, 1940.

Davis, Gherardi. *The United States Navy and Merchant Marine from 1840 to 1880.* New York, 1923.
   Brief survey and reminiscences by a career Navy man.

Feipel, Louis N. "The United States Navy in Mexico, 1821-1914: Chapter V, War with Mexico—Operations on the East Coast, 1845-1848," *United States Naval Institute Proceedings,* Vol. XLI (July-August, 1915), 1159-72.

———. "The United States Navy in Mexico, 1821-1914: Chapter VI, War with Mexico—Operations on the East Coast," *United States Naval Institute Proceedings,* Vol. XLI (September-October, 1915), 1527-34.

Hayes, John D. "Loss of the Norfolk Yard," *Ordnance,* Vol. XLVI (September-October, 1961), 220-23.
   Brief but scholarly account, written by a naval historian currently working on the letters of Rear Admiral S. F. du Pont.

Hibben, Henry B. *Navy Yard, Washington: History from Organization, 1799 to Present Date.* Washington, D. C., 1890.

Jane, Fred T. *The British Battle Fleet: Its Inception and Growth throughout the Centuries to the Present Day.* 2 vols. London, 1915.
   Volume one briefly discusses the influence of the *Wampanoag* on the *Inconstant*-type of warship.

Johnson, Robert Underwood, and Clarence Clough Buel, eds. *Battles and Leaders of the Civil War.* 4 vols. New York, 1884-87.

Long, John D. *The New American Navy.* Vol. I. New York, 1903.

Long, John Sherman. "The Gosport Affair, 1861," *The Journal of Southern History,* Vol. XXIII (May, 1957), 155-72.
   A thorough, scholarly account based on published government documents and a few manuscript collections. Little mention of Isherwood.

Lull, Edward P. *History of the United States Navy-Yard at Gosport, Virginia.* Washington, D. C., 1874.
   See chapter ten for the loss of the yard in April, 1861. Quotes a number of official reports, including Isherwood's.

Marvin, Winthrop L. *The American Merchant Marine: Its History and Romance from 1620 to 1902.* New York, 1902.

Merrill, James M. "Strategy Makers in the Union Navy Department, 1861-1865," *Mid-America: An Historical Review,* Vol. XLIV (January, 1962), 19-32.

Paullin, Charles Oscar. "Early Voyages of American Naval Vessels to the Orient: XIII—The East India Squadron in the Waters of China and Japan, 1854-1865," *United States Naval Institute Proceedings,* Vol. XXXVII (June, 1911), 387-417.

———. "A Half Century of Naval Administration in America, 1861-1911: I—The Navy Department during the Civil War, 1861-1865," *United States Naval Institute Proceedings,* Vol. XXXVIII (December, 1912), 1309-36.

———. "A Half Century of Naval Administration in America, 1861–1911: II—The Navy during the Civil War, 1861–1865," *United States Naval Institute Proceedings,* Vol. XXXIX (March, 1913), 165–95.

———. "A Half Century of Naval Administration in America, 1861–1911: III—The Second Administration of Gideon Welles, 1865–1869," *United States Naval Institute Proceedings,* Vol. XXXIX (June, 1913), 735–47.

———. "A Half Century of Naval Administration in America, 1861–1911: IV—The Navy Department under Grant and Hayes, 1869–1881," *United States Naval Institute Proceedings,* Vol. XXXIX (June, 1913), 747–60.

———. "A Half Century of Naval Administration in America, 1861–1911: VIII—The New Navy, 1881–1897," *United States Naval Institute Proceedings,* Vol. XXXIX (December, 1913), 1469–87.

Scholarly and perceptive accounts which have been modified only in minor details by subsequent research.

———. "Naval Administration, 1842–1861," *United States Naval Institute Proceedings,* Vol. XXXIII (December, 1907), 1435–77.

———. "President Lincoln and the Navy," *American Historical Review,* Vol. XIV (January, 1909), 284–303.

Nothing on Isherwood, but good on the general relationship between Lincoln, Welles, and Fox.

Penn, Geoffrey. "Up Funnel, Down Screw!" *The Story of the Naval Engineer.* London, 1955.

A history of the introduction of steam into the British Navy. Excellent for the attitude of line officers toward the encroachment of the engineers.

Potter, E. B., and Chester W. Nimitz, eds. *Sea Power: A Naval History.* Englewood Cliffs, New Jersey, 1960.

Comprehensive textbook currently being used at the United States Naval Academy. Heavily weighted toward the twentieth century, but good for the brief survey of technological progress of navies during the nineteenth century.

Pratt, Fletcher. *The Navy, A History: The Story of a Service in Action.* New York, 1941.

Relentlessly over-dramatized account of the development of the American Navy. Particularly complimentary toward Isherwood, "the man who really won the naval war," and the *Wampanoag.* Draws largely from Bennett for these subjects.

Sprout, Harold, and Margaret Sprout. *The Rise of American Naval Power, 1776–1918.* Princeton, 1939.

The best history of American naval policy up to 1880. Briefly discusses Isherwood's and Lenthall's theories of sea power.

Westcott, Allan, ed. *American Sea Power since 1775.* Philadelphia, 1947.

# Index

ABCD ships: 239
Albany Academy: reputation in 1830's, 5; curriculum, content, and purpose of, 6; compared with colleges, 6
Alden, James, naval officer: 22, 183, 185–86, 201; rescue of *Merrimack*, 23–24, 25–26; criticizes *Wampanoag*, 184–85
*Algonquin:* 141, 256; description, 119, 126; contractual provisions for, 120, 254; dock trials, 120–21, 123, 124–27; sea trial, 123, 128–29; engine performance, 129; *see also Winooski*
*Allegheny:* 14
Allen, Horatio: 91
American Society of Mechanical Engineers: 241–42
American Society of Naval Engineers: 241
Ammen, Daniel, naval officer: 203, 229
*Ammonoosuc:* 145, 170, 173, 256; speed trial, 186–87; engine efficiency, 187; *see also Wampanoag*
"Anaconda Policy": 27
*Antietam:* 32
*Alabama:* 32, 187
Archbold, Samuel, naval engineer: 20
*Arizona:* 180
Armstrong, James, naval officer: 16–17
*Army and Navy Journal:* 143–44; *see also* Church, William C.
Arthur, Chester A.: 237, 239

Beck, Dr. T. Romeyn: 7
Bell, Henry H., naval officer: 16

Bennett, Frank M., naval engineer: 115, 120
Blockade of Confederacy: 27; Isherwood's views on, 246
Board of Admiralty: 136; 1865 proposal, 198; 1867 proposal, 199, 202; 1869 proposal by Grimes, 225; Welles's criticism of, 198; *see also* Porter, David Dixon
Board of Survey: *see* Board of Admiralty
Boiler, Martin water-tube: *see* Martin water-tube boiler
Boilers, steam, and relation to engine size: 93–94; *see also* steam pressure
Borie, Adolph: 233, 234; as figurehead in Navy Department, 229, 231; reluctant to remove Isherwood, 229
Brassey, Sir Thomas: 188, 249
Browning, Senator Orville H.: 134
Bureau system, Navy Department: inefficiencies of, 197
Bureaus, Navy Department: relations with Congress, 197–98; *see also* bureaus by name

Chandler, William E.: and Zeller tool purchases, 216; and ABCD ships, 239
*Charleston:* 180
*Chattanooga:* 145, 148, 179; sea trial, 168–69
*Chimo:* 71–72
Church, William Conant: supports Ericsson, 143–44; depends on Ericsson for engineering advice, 144; supports monitors, 144; aligned with naval reactionaries, 144; on *Wampanoag's* speed trial,

287

182, 260; supports Board of Admiralty, 199; biography of Ericsson, 256
Coal: Isherwood's experiments with, 82, 234; postwar economizing on use of, 159–60
Coles, Cowper Phipps, English naval officer: 50; *see also* turrets, Coles's
Colfax, Schuyler: 198
Committee on the Conduct of the War: 74–75
Compound engine: 251
Condenser: *see* Sewell condenser
*Constellation:* 44
*Constitution:* 32, 54
Construction, Equipment, and Repairs, Bureau of (to July, 1862) 43; relation to office of engineer in chief, 4, 28; *see also* Lenthall, John
Construction and Repair, Bureau of: 44; and light-draft-monitor contracts, 68; budget for 1864 and 1865 fiscal years, 133; criticized by Isherwood in the 1870's, 237; *see also* Lenthall, John
*Contoocook:* 32, 161
Contractors, civilian: deficiencies in wartime construction, 37–38, 247
Coryell, Miers, marine engineer: 123
Cramp, Charles H., shipbuilder: 67; *see also Chattanooga*
Craven, Thomas T., naval officer: 235
*Cumberland:* 56
Cutoffs: *see* Sickels' cutoff; steam cutoffs

*Dacotah:* 20
Dahlgren, John, naval officer: 60
Danby, Robert, naval engineer: 22–23
Daniel, R. P., naval surgeon: treats Isherwood's dysentery, 18–19
David W. Taylor Model Basin: 260
"Davidoff Gunboat": 30; *see also* ninety-day gunboat
Davis, Representative Henry Winter: 114, 136; and 1864 investigation of Isherwood's machinery, 137; and Board of Admiralty, 198
Delano, B. F., naval architect: 145; designs *Wampanoag* with Isherwood, 170–72
Dickerson, Edward Nicoll: 140, 141; and Stimers, 71; education and training of, 105; description, 105; reputation as patent lawyer, 105; and Isherwood's 1861 nomination, 108, 111; pamphlets attacking Isherwood, 108–10, 111–12; reasons for opposing Isherwood, 108; cross-examination of Isherwood, 110, 111; attacks *Experimental Researches,* 111–12; accuses Fox of persecution, 115–16; forcing congressional investigation of Isherwood's machinery, 133–34, 137–38; theories of steam expansion, 139; *see also* steam expansion, *Navy of the United States* (pamphlet), *Steam Navy of the United States* (pamphlet)
Dickerson, Edward Nicoll, engine designs of: for Detroit waterworks, 106; for *Pensacola,* 106–107; criticized by Isherwood, 107; lampooned in pamphlet, 115; for *Idaho,* 117; for *Algonquin,* 119, 124, 126, 129
*Dictator:* 60, 62
Doolitte, Senator James R.: 255–56
Double-ender gunboats: 31, 40, 119, 247; speed of, 31, 32, 139; *see also Algonquin* and *Winooski*
Draft of skilled workers: 38–39, 40
*Dreadnought* (British): 56
*Dunderberg:* 63–64
Du Pont, Samuel F., naval officer: 70, 71; supports Board of Admiralty, 198
Dysentery, chronic: Isherwood ill with, 16, 18–19, 236; treatment of, in 1850's, 18; seriousness of disease in 1850's, 18–19

Eads, James B.: 63

Economic retrenchment, postwar, effect on Navy Department and Isherwood's bureau operations: 159–60, 166
Efficiency, marine engine: 93, 129, 138, 155, 179, 187, 246, 247, 251
Engineer in chief, office of: attraction for civilians, 3; within construction bureau, 4, 28; staff in 1861, 29; wartime problems in retaining staff, 42–43; relationship to chief of steam bureau, 44; attempts to put in civilian, 44, 189, 202, 222–25, 261; Isherwood's 1883 request for appointment, 239; see also Steam Engineering, Bureau of
Engineering, American, compared with British: 136–37, 140
*Engineering Precedents for Steam Machinery*: 82, 83, 108, 109
Engineers, marine (civilian): relation to naval engineers in 1860's, 3
Engineers, naval: see naval engineers
Engineers, steam, lack of education in mid-nineteenth century, 80
Engines, geared: see geared engines
Engines, marine: in early 1860's, 80; see also efficiency, marine engine; geared engines; Isherwood, engine designs of; steam expansion
English engineers, views of: on *Experimental Researches*, 95–96; on *Algonquin-Winooski* trials, 130, 254–55; on Isherwood, 155; on *Wampanoag* trial, 182–83
Equipment and Recruiting, Bureau of: 44
Ericsson, John: 26, 49, 58, 62, 141; and *Princeton*, 10; opposes Isherwood over monitors, 51, 67–68; competing with Coles turret, 51; monitor designs, 60; designs light-draft monitors, 66; relinquishes direction of light-draft-monitor program, 68; repudiates light-draft-monitor plans, 69, 72; breaks with Stimers, 70; relationship with Fox, 76; and steam-expansion experiment board, 91; engine designs of, 93, 145, 146, 153, 156, 247; supports Dickerson, 110; and *Algonquin-Winooski* trials, 126; characteristics as an inventor, 143; engineering advisor for *Army and Navy Journal*, 144, 154; opposes Isherwood's theories, 145; business relationship with Isherwood, 146, 147; opinion of Isherwood, 149–50; opposes Isherwood's renomination as bureau chief, 152; and *Wampanoag* trial, 181–82; extolled by Porter, 203, 205; and ventilation system on *Monitor*, 220–21; connection with English engineering journals, 222; biography by Church, 256; and Isherwood's boiler system, 258
Erie experiments: see *Michigan*
*Eutaw*: 138, 139
Expansion of steam: see steam expansion
Experiment boards: Isherwood's service on, after 1876, 236–37
*Experimental Researches in Steam Engineering*: 80, 92–96, 97, 111–12, 145, 163–164

Farragut, David G., naval officer: 107, 174, 240; critical of wartime engineers, 34; opposes Isherwood's views on steamer design, 167; relationship to Porter, 202
Faxon, William: 229
Fire-tube boiler: 82, 83, 139
Fisher, Clark, naval engineer: 29
Fleet engineer: 189–90
*Florida*: see *Wampanoag*
Forbes, Paul S.: backs Dickerson's projects, 116; opposition to Isherwood, 117; offers to build *Idaho*, 117; and construction of *Idaho*, 117–18; financial problems in ship construction, 118; involved in Isherwood-Dickerson dispute,

121–22; and *Idaho* contract, 131, 253; and *Algonquin* contract, 254
Forced-draft system for boilers: 31, 129, 247
Forney, John N.: 204–205
Fox, Gustavus Vasa: critical of delays in ship construction, 43; views on ironclads, 50, 66; supports Ericsson's monitor designs, 51, 62–63; and Eads's monitor, 62–63; wartime role in Navy Department, 76–77; ridiculing Dickerson, 115; supports Isherwood's experiments, 164, 180; postwar political influence, 227–28, 229–30; correspondence with Isherwood, 259
Franklin Institute: 91, 92, 95
Franklin Institute *Journal*: see *Journal of the Franklin Institute*

Geared engines: 39, 59, 145, 171–72, 173, 176, 178, 181, 259
*General Taylor*: 9
*Gloire* (French): 49
Goldsborough, John S., naval officer: 235
Goldsborough, Louis M., naval officer, and 1869 board investigating Isherwood's machinery: 230, 233–34
Gosport Navy Yard, Norfolk, Virginia: 21, 25–26
Government iron and armor manufacturing facilities: 52, 219
Government machine shops, wartime need for: 42
Grant, Ulysses S.: 204, 205, 228, 233, 237; and Porter, 225–26, 264; removes Isherwood from office, 231
Greeley, Horace: 136
Green, John, civil engineer: 8
Gregory, Francis H., naval officer: 174; head of "monitor bureau," 67; relations with Stimers, 68, 70; and *Algonquin-Winooski* trials, 124, 125; and *Madawaska* trial, 153
Grimes, Senator James W.: 189, 223, 225, 249; and 1862 Navy Department reorganization, 44; and Isherwood's nomination as steam bureau chief, 45; on ironclads, 49–50; on Isherwood's and Lenthall's ironclad proposals, 64–65; and Dickerson, 116; defends Navy Department, 136; and Isherwood's renomination as bureau chief, 152; opinion of Lenthall, 165; hostility toward Isherwood, 200; supports line officers, 200; influenced by Porter, 206–207; introduces Board of Admiralty bill, 225; disillusioned with Porter and future of Navy Department, 226
Griswold, John A.: on Forbes and *Idaho*, 118; relationship with Ericsson, 147; and Ericsson-Isherwood relationship, 147–49; see also Ericsson, John; *Madawaska*
*Guerriere*: 32

Hale, Senator John P.: 114, 136, 249, 255; and Isherwood's 1861 nomination, 4; attacks Navy Department, 135
Hampton Roads, Battle of: see *Monitor, Virginia*
Hanscom, Isaiah: 237
Harris, Townsend: 16–18
*Hartford*: 174
Hartt, Edward: 57
Haswell, Charles, naval engineer: 12; reorganizes engineer corps, 9; modifies machinery of *Princeton*, 10; failures with *San Jacinto*, 12–13; removed from office, 13; criticized by Isherwood, 13, 82
Heat, theories of, in 1850's: 79–80
Heat loss, in steam engines: 85–86; see also steam engines, insulating of
Henry, Joseph: at Albany Academy, 5; opinion of Albany Academy curriculum, 6; and Dickerson, 105, 109
Hong Kong, China: headquarters of East India Squadron, 17, 18

*Index* 291

Hunt, William H.: 238
Hunter paddle wheel: 14

Idaho: 119, 145, 148, 168, 179, 255; Forbes's offer to build, 117; construction of, 118; sea trial, 130; as a sailing vessel, 131; *see also* Dickerson, Edward Nicoll
*Inconstant* (British): 183, 188; progenitor of modern battle cruiser, 261
Indicator diagrams: 250
Inflation, wartime problems with: 34, 35, 38, 39, 40, 62
Insulation: *see* steam engines, insulating of
Ironclads: armor of, 50–51, 52, 53, 55, 56, 57, 58, 59, 60, 61, 63, 64, 66, 248; 1862 board of, 57; casemated, 59, 61, 62 63, 64, 144, 248–49; designed by Isherwood and Lenthall, 50, 51, 59, 60, 61, 62, 134; harbor defense, 50, 51–52, 54, 57–58; Isherwood's and Lenthall's views on proper use of, 54; technological limitations of, in 1860's, 54–55; *see also Monitor;* "monitor bureau"; monitors; *New Ironsides; Virginia*
Ironclads, light-draft: *see* light-draft monitors
Ironclads, monitor; rejected by Isherwood and Lenthall, 52–53; *see also* monitors
Ironclads, seagoing: 49–50, 53, 54, 56, 58, 59, 60, 61, 62, 63, 64, 65, 144, 238; supported by Welles, 55
Ironclads, turreted: 37, 50–51, 56, 57, 58, 63, 64, 144; *see also* monitors; turrets
Isherwood, Benjamin Franklin: appointed engineer in chief in 1861, 4; nomination as engineer in chief opposed in Senate, 4; public reception to nomination, 4–5

Birth and parents, 5; relationship with mother, 5; enrolled in Albany Academy, 5; preparation in mathematics, 6; scholastic record, 7; expelled in 1836 from Albany Academy, 7; employment with railroads, 7; early training in steam machinery principles, 7, 81; first publications, 7–8; as a civil engineer, 8; and lighthouse construction and lense designs, 8; training on marine engines, 9, 81

Appointed as first assistant engineer in Navy, 9; duty at Pensacola, Florida, 9; examined in 1845 for re-ranking, 9–10; demoted in rank and seniority, 10; naval service in Mexican War, 10–12; duty on *Princeton*, 10–11; as senior engineer on *Spitfire*, 11–12; commended by Tatnall and Porter, 12; promoted to first assistant engineer, 12; duty in office of engineer in chief, 12; lighthouse duty, 12, 14; promoted to chief engineer, 12; becomes commissioned officer, 12; early reports in steam engineering, 12; redesigns propeller and rudder of *San Jacinto*, 13; made assistant to engineer in chief, 13; attacks Haswell's machinery designs, 13, 82; designs propeller and rudder, 13; early contributions to *Journal of the Franklin Institute*, 14; designs back-acting engine, 14; problems with *Allegheny* machinery design, 14; designs feathering paddle wheel, 15; and 1852 petition to increase size of engineer corps, 15–16; sea duty on *Massachusetts*, 16; ill with dysentery, 1854, 16; as chief engineer of *San Jacinto*, 16–18; compiling vessel and machinery data, 17, 19, 82; attack of dysentery in Far East, 18–19; escapes being poisoned in Hong Kong, 18; on experiment boards in 1859–61, 19, 20; on 1860 board of survey for conversion to steam warships, 19; steam expansion experiments on *Michigan*, 19–20, 87–90; prewar reputation as a naval engineer, 20; friendship with Zeller, 20, 235, 236, 240, 241;

friendship with Lenthall, 20, 28, 164–66

Attempts to repair *Merrimack*, 22–23; contemplates rescuing *Merrimack*, 24–25; escapes kidnapping at Norfolk, 25; evaluation of his role at Gosport Navy Yard, 26

Physical appearance, 27–28; personality characteristics, 28, 143, 227, 231–32, 241; relations with office staff, 29; and purchase of steamers during wartime, 35; contracting for construction of naval machinery, 35–42; antagonism toward private constructors in wartime, 41–43; urges government manufacturing and repair facilities, 41, 42, 52, 219; wartime duties summarized, 47

Nominated in 1862 as chief of steam bureau, 44, 45, 46, 144–45; salary and tenure as chief, 44; receives commission, 45; appointment confirmed, 46; rank as chief, 46; in creating bureau, 189, 261; powers as chief, 197–98; status as chief, 263

Said to oppose ironclads, 50, 147; designs ironclads with Lenthall, 50, 51, 59–60, 61, 62; doctrine of seapower (with Lenthall) in 1862, 52–54; and conversion of *Roanoake* into ironclad, 55–56; serves on 1862 ironclad board, 57–58; said to oppose monitors, 66; role in light-draft-monitor episode, 67, 73–74, 77; and machinery of light-draft monitors, 69; testimony on light-draft monitors, 75; evaluation of Stimers, 75

And rise of engineering profession, 79; an engineer before a naval officer, 79; and study of thermodynamics, 81; articles in *Journal of the Franklin Institute*, 81–82; writes *Engineering Precedents*, 82; endorses Martin boiler, 82–83; on limitations of steam expansion, 83–86; experiments to disprove Mariotte's Law, 84–86, 87–90; views on experimental method, 86–87, 96; on weakness of contemporary scientfic method, 86, 87, 89, 97–102; on the inductive process of experimentation, 89, 97–98; experiments evaluated by Rankine, 89–90; work evaluated by Thurston, 90, 96; experiments with superheated steam, 90–91; and steam expansion experiment board, 91; experiments with petroleum as engine fuel, 91–92, 162–63, 252; and publication of *Experimental Researches*, 92–96; views on relation between boiler and engine size, 93–94; on role of engineering in society, 97; interpretation of Civil War, 97, 209; philosophy of engineering, 96–103; as an engineering iconoclast, 99–101; reaction to criticism, 101–102, 115, 120, 122–23, 154, 222

On Dickerson's machinery in *Pensacola*, 107; source of his wealth, 112–13, 200, 206, 240–41; and *Algonquin-Winooski* trials, 120, 124, 126–27, 130; and 1863 civilian investigating board, 134; supported by Donald McKay, 136–37; supported by congressional committee, 138–41; opposed by Church, 144; negotiations over *Madawaska* contract, 146–51, 153–54, 157

Nomination in 1866 as bureau chief, 151–52; tenure in office, 230, 265; confirmation of appointment, 264–65; postwar cutbacks in machinery testing, 160–61; engine sent to Naval Academy, 162, 191; criticizing civilian marine engineers, 162; on primacy of speed in steamer design, 166–67; and design of *Wampanoag*, 171–73, 185; supervising *Wampanoag*'s construction, 173, 259; and *Wampanoag*'s trial, 180–81; criticizes Alden on *Wampanoag*, 185–86; opinion of *Army and Navy Journal,*, 186

And fleet engineer, 189–90; promotes education of engineers at

Naval Academy, 190-91; proposal to reorganize engineer corps, 192; urging higher pay and position for naval engineers, 192-93, 209; ranked with commodores, 193; as target of Board of Admiralty proponents, 199-201; supports Moderate Republicans, 204; political affiliations, 204, 239, 240; 1867 proposal for higher rank and pay for engineers, 208-209; on predominant role of engineers in Navy, 208-209; rank as rear admiral, 209-11, 240; and Zeller tool purchases, 213-19; explanation of tool purchase affair, 214-15; postwar machinery construction, 219, 228; and ventilation of *Monitor,* 220-21; seeks support from Welles to stay in office, 228; solicits Fox's help to remain in office, 229-30; removed from office of bureau chief, 231; evaluated by Welles, 231-32; ordered to leave Navy Department, 232

Coal and propeller experiments at Mare Island, California, 234; relative rank reduced in 1869, 235; and line-officer hostility at Mare Island, 235; fitness evaluation at Mare Island, 235; as consulting engineer at Key West, Florida, 235; recurrence of dysentery in 1873, 236; ordered in 1875 to Japan, 236; European tour with Zeller, 236; as critic with Lenthall of Navy Department in the 1870's, 237; minority report with Lenthall of First Naval Advisory Board, 238-39; critic of Navy in the 1880's, 239; opinion of ABCD ships, 239; effort to become steam bureau chief before retirement, 239; engineering advisor to Porter, 240; retires from Navy, 233, 240

Suggested as candidate for engineer in chief in 1885, 240; residence in New York city, 240-41; life in retirement, 241; contributor to *Journal of American Society of Naval Engineers,* 241; made honorary member of American Society of Mechanical Engineers, 241; testimonial dinner in 1898, 242; bronze bust of, 242; loses fortune, 242; honored in 1915 by Naval Academy, 242; dies, 242

Views on blockading the Confederacy, 246; work with compound engine, 251; influence on scientific method and engineering training, 252; postwar correspondence with Fox, 259

Isherwood, Benjamin Franklin, engine designs of: 93-94, 246, 247; criticized, 14-15, 60, 95-96, 101, 126, 133-34, 137, 140, 145, 146, 155, 167-68, 174-75, 176, 177, 181-83, 185, 222, 223-24, 233, 237, 247, 258, 259; for "Davidoff Gunboat," 30; for double-ender gunboats, 31, 32, 247; "Isherwood engine," 14; for ninety-day gunboats, 31, 246; for screw sloops of war, 31, 32, 247; for large steam frigates, 32; total for 1861-65 period, 33; emphasis on durability and reliability, 33-34, 140, 247; for ironclad cruiser, 59, 61, 62; investigation of, in 1863, 60, 111; for *Wampanoag,* 117, 171-72; for *Winooski,* 119; congressional investigation of, in 1864, 137-41; compared with British and French machinery, 140, 161-62

Isherwood, Eliza: 5, 112-13, 236
"Isherwood engine": 14
Isherwood Hall: 242
Isherwood wheel: *see* paddle wheel, feathering
*Itasca:* 139

Jane, Fred T.: 261
Japan: 16, 18
Jenkins, Thornton, naval officer: 183-84
Johnson, Andrew: 152, 204, 219, 226

*Journal of the Franklin Institute:* 14, 81–82, 86
*Juniata:* 37

Kansas: 161
*Kearsarge:* 31
Kelley, Representative William D.: investigates Zeller tool purchases, 214, 215–18; criticizes tool purchases, 217; investigates Isherwood's bureau operations, 218–19; supports bill to put in civilian as steam bureau chief, 222; criticizes American naval engineering, 223–24; conduct of his investigating subcommittee, 263
Kerwin, Madeleine: 113
King, James W., naval engineer: and 1852 petition to increase size of corps, 15; supervises gunboat construction, 70–71; description, 226–27; early opposition to Isherwood, 226; solicits Fox's aid in replacing Isherwood, 227–28; nominated and confirmed as steam bureau chief, 231; on Mariotte's Law, 251; relationship to tool-purchase investigations, 264

Labor, wartime scarcity of: 38–39, 40, 41, 118
Lake, William, civil engineer: 7
Latimer, W. K., naval officer: 9; commends Isherwood, 10
Lenthall, John: 21, 30, 43, 47, 235, 236; and friendship with Isherwood, 20, 28, 164–66; designs *Merrimack*, 21; personality, 28; nominated as chief of Bureau of Construction and Repair, 44, 45, 46–47; designs ironclads with Isherwood, 50, 51, 59–60, 61, 62; said to oppose ironclads, 50, 147; doctrine of seapower (with Isherwood) in 1862, 52–54; on 1862 ironclad board, 57–58; said to oppose monitors, 66; and light-draft monitors, 73–74; evaluation of Stimers, 77; opposes Dickerson in 1858, 106; views on Ericsson, 147; serious illness, 164–66; powers as bureau chief, 197–98; critic of Navy Department in 1870's, 237; and minority report of First Naval Advisory Board, 238–39
Light-draft monitors: 205, 220; need for, 65–66; Ericsson's design, 66; Ericsson's plan turned over to "monitor bureau," 67; failure of design, 71–72; congressional investigation of, 74–75
Lincoln, Abraham: 190; nominates Isherwood in 1861, 3–4; and Gosport Navy Yard crisis, 21, 25, 26; and formal blockade of the Confederacy, 27; nominates Isherwood as Steam Bureau Chief, 44, 45, 46
Line officers: and Isherwood, 24, 26; hostility toward Isherwood and Lenthall, 45–46, 193 ff.; *see also* Porter, David Dixon
Line-staff controversy: reasons for, 193–99, 208–11
Long, Robert, naval engineer: 19

*Madawaska:* 145, 169, 170, 179, 256; contractual provisions for, 146, 149; engines of, 146; progress payments on, 148, 149; preparations for trials, 150–51; preliminary trials, 153, 154; speed of, 154, 155; final sea trial, 155–56; engine efficiency, 155; design problems, 156; contract payments, 157; construction of, 173; lack of durability explained, 258; *see also* Ericsson, John
Mahan, Alfred Thayer: seapower doctrine related to views of Isherwood and Lenthall, 53, 54; on warship names, 257
Mallory, Senator Stephen: 45, 106
*Marblehead:* 191
Mare Island Navy Yard, California: 234–35
Mariotte's Law: 86, 88, 98, 100, 106, 109, 139, 251; described, 84; *see also* steam expansion
Martin, Daniel, naval engineer: 57,

60, 82, 106, 108, 112
Martin water-tube boiler: 60, 108, 112, 119, 133, 134, 139, 172, 175, 185, 186, 246, 247; endorsed by Isherwood 82–83
*Massachusetts:* 16, 176
McCauley, Charles E., naval officer: 21, 22; refuses to release *Merrimack,* 23–24; scuttles *Merrimack,* 25
McKay, Donald, shipbuilder: 136–37
McKay, Nathaniel, shipbuilder: 240
Melville, Herman: 7
*Merrimack:* 45, 55, 93; repairing at Gosport Yard, 21–23; attempted rescue of, 23–25; scuttled, 25; consequences of her loss to the Confederacy, 26; value of, 246; *see also Virginia*
*Miantonomah:* 51, 256
*Michigan:* 19–20, 87–91
*Monadnock:* 51
*Monitor:* 26, 49, 51, 52, 58, 62, 67, 93, 143, 147, 201, 249; influence on Navy Department policy toward ironclads, 57, 66; ventilation of, 220–21
"Monitor bureau": 73; established in 1861, 66–67; autonomous in Navy Department, 67
Monitors: 64, 144, 238; and American naval strategy, 26, 248; designed by Isherwood and Lenthall, 51; designed by Eads, 63; criticized by English engineers, 221–22; *see also* Ericsson, John; ironclads
Monitors, light-draft: *see* light-draft monitors
Morgan, George D.: 35

National Academy of Sciences: 91, 139
Naval Academy, United States: Isherwood's engines at, 162, 191; engineering education at, 190–91; establishes steam engineering department, 191; Porter as superintendent, 202–203, 206; names Isherwood Hall, 242

Naval Advisory Board, First (1881): 238–39
Naval Advisory Board, Second (1882): 239
Naval contractors: and machinery construction during wartime, 34–35
Naval engineers: status in Navy in 1860's, 3; wartime problems with inexperience, 33–34; as commissioned officers, 12, 193; reranked in 1863, 193, 195; inferior status of, 193–96; sources of conflict with line officers, 193–98, 208–11
Naval engineers, board of: investigating Zeller tool purchases, 215, 216; disbanded, 230
Naval Engineers, Corps of: created, 8–9; petition to increase size, 15–16; improvement of, 34; 1865 reorganization proposed, 192; postwar resignations from, 209
Naval engineers, training of: part of Naval Academy curriculum, 190–91
Naval machinery, contract negotiation procedure during wartime: 36–37
*Navy of the United States, The* (pamphlet): 71, 110–13, 115, 138; *see also* Dickerson, Edward Nicoll
Navy, United States, vessel status in March, 1861: 27, 246
Navy Department, United States: appearance in 1861, 29; purchase of steamers during wartime, 35; reorganization in 1862, 44; support of Isherwood's publications, 94–95; budget for 1864 and 1865 fiscal years, 133; wartime criticism of, 134–36; proposed reorganization in 1865, 198; wartime problems with contractors, 247
Navy Department bureaus, United States: *see* bureaus by name
*Neshimany:* 145, 170
*New Ironsides:* 59, 61, 248–49
Newton, Isaac, naval engineer: 154, 208; supports Dickerson, 110; and *Madawaska* trials, 153; hired

by line-officer association, 201, 220–21; and light-draft monitors, 220; opposes Isherwood, 220–21; on Goldsborough Board in 1869, 233

Nicholson, J. W. A., naval officer: 176, 177, 178, 184; and *Wampanoag* trial, 177–78

Ninety-day gunboats: 31, 181, 246, 247; engines of, 247

*Nipsic:* 32, 139, 161

Novelty Iron Works, New York city: 30, 31, 91; employs Isherwood, 9; converts *Roanoake* to ironclad, 56; builds engines of *Wampanoag*, 173, 174

Norfolk Navy Yard, Virginia: *see* Gosport Navy Yard

Nye, Senator James W.: 206; supports Board of Admiralty, 199, 201, 206, 211; attacks engineers, 200–201

Oil: *see* petroleum
*Ossipee:* 134

Paddle wheel, feathering: 15
Parrott, Enoch G., naval officer: 235
*Passaic:* 93
Paulding, Hiram, naval officer: 25, 174
*Pensacola:* 106–107, 116, 134, 137, 168
Petroleum, as engine fuel: 91–92, 162–63, 252
*Pompanoosuc:* 170, 256
Porter, David Dixon: executive officer of *Spitfire*, 11–12; relations with Isherwood during Mexican War, 11–12; criticism of Lenthall and Isherwood, 45–46; recollections of Isherwood in Mexican War, 46; on Isherwood engines at Naval Academy, 162; characteristics, 202–203; as Naval Academy superintendent, 202–203, 206; and postwar politics, 203; views on naval engineers, 203, 205; opinions of Isherwood, 203–204; sides with Radical Republicans, 204; newspaper campaign against Isherwood and the engineers, 204–206; promoting Board of Admiralty scheme, 204, 205, 206, 207, 210–11; reaction to Isherwood's proposal for high rank, 210–11; support of Van Horn's bill, 223, 224–25; attitude toward Grant, 225–26, 264; takes over Navy Department, 229, 230–31; appoints Goldsborough Board to investigate Isherwood's bureau operations, 230, 233; in control of American Navy, 233–34; reconciliation with Isherwood in later years, 240; recommends Isherwood as steam bureau chief in 1881, 240

*Princeton:* 10–11
Progress payments to contractors: 36, 117, 118, 147, 148, 149
Propeller experiments: at Mare Island by Isherwood, 234

*Quinnebaug:* 161–62
Quintard, George, shipbuilder: 118

Rams, naval: 53, 55, 56, 63, 64, 116, 172, 238
Randall, Representative Samuel J.: 218
Rank: differentiated from command, 194; controversy over, 194–95, 208–11; of bureau chiefs, 236
Rank, relative: of engineers, 193, 195, 209, 235
Rankine, W. J. Macquorn: 80, 81, 89–90
Raymond, Henry: 136
Reed, Sir Edward J.: 183, 188
Republicans, Radical vs. Moderate: 204, 218, 219, 220, 264
Rice, Representative Alexander H.: 118, 138, 198
*Richmond:* 107–108, 166
Roach, John, shipbuilder: 219, 263, 264; sale of machine tools to government, 213–19; builder of ABCD ships, 239; *see also* Zeller tool purchases

*Roanoake:* 37, 55, 59, 93; as predecessor of modern battleship, 56–57
Robeson, George: 234; removes Isherwood from Mare Island, 235; accuses Isherwood of being tool of Democratic party, 237
Robie, E. D., naval engineer: 255
Rochambeau (French): *see Dunderberg*
Rowbotham, John: 213–14, 215

*San Jacinto:* failure of original design, 13; cruise to Far East, 16–19
*Sassacus:* 31, 40, 119
Seapower, Isherwood's and Lenthall's views on: 52–54
Secretary of the Navy: *see* Toucey, Welles, Borie, Robeson, Whitney
Select Committee on Naval Supplies: 135, 255–56
Selfridge, Thomas O., naval officer: 214
Seward, William H.: 136
Sewell, William, naval engineer: 108
Sewell condenser: 108, 112, 133, 140, 246
Siam: and American trade treaty, 16; visit by Townsend Harris and American naval officers of *San Jacinto*, 17
Sickels, Frederick, and Dickerson in 1850's: 106–107
Sickels' cutoff: 107, 110, 116, 119; *see also* steam cutoffs
Simon's Bay, South Africa: Isherwood's visit on *San Jacinto*, 17, 82
Skinner, Charles: 13
Smith, Joseph, naval officer: 57; and light-draft-monitor designs, 69
Smith, Melancthon, naval officer: 183–84
Speed: of steam warships; 64, 107, 127, 129, 130, 134, 155, 161–62, 166, 168, 178, 179, 180, 183, 184, 187, 246, 248; of steamers with Isherwood's machinery, 129, 138–39, 161–62, 166, 178, 179, 180, 184, 246; of prewar steamers, 138
*Spitfire:* 11–12
Stanton, Edwin: 135
Steam, superheated: 89, 90–91, 93, 124, 138, 150, 151, 172, 175, 252
Steam condensation: 85–86, 90, 93, 145, 251
Steam cutoffs: 84, 87, 88, 91, 94, 119, 129, 139, 161, 162, 246, 247; *see also* Sickels' cutoff
Steam-engine designs: weakness of, mid-nineteenth century, 80–81
Steam engineering: limitations on development, 79–81; status at mid-nineteenth century, 79–81
Steam Engineering, Bureau of: 28; created in 1862, 44; office staff in 1862, 44; budget for 1864 and 1865 fiscal years, 133; budget for postwar fiscal years, 159; postwar reduction of work force, 160; Isherwood's role in creating, 189, 261; criticized by Isherwood in 1870's, 237; *see also* engineer in chief, office of
Steam engines, efficiency of: *see* efficiency, marine engines
Steam engines, insulating of: 80, 85, 88, 89, 93, 251; *see also* heat loss
Steam expansion: 106, 119, 122, 127, 131, 134, 139, 145, 161, 251; experiments on *Michigan* in 1860–61, 19–20, 87–91; limits in engine design, 33; *see also* Mariotte's Law
Steam expansion, theory of: 83–84
Steam-jacketing: *see* steam engine, insulating of
Steam pressure: in 1860, 88–89; *see also* Martin water-tube boiler
*Steam Navy of the United States, The* (pamphlet): 108–10; *see also* Dickerson, Edward Nicoll
Steam warships: living conditions on, in 1840's, 11; in commission at outbreak of Civil War, 246
Steaming reports, monthly: 159–60
Steel, use of: opposed by Isherwood and Lenthall for warship construction, 238–39
Steel industry, American: 239

Steers, Henry, ship designer: 117, 145
Stevens, Thaddeus: 198
Stimers, Alban, naval engineer: 19, 108; appointed general inspector of ironclads, 67; relations with Isherwood, 67, 74, 75; establishes own monitor office, 68; dispute with Admiral Du Pont over ironclads, 70, 71; attempts to be made engineer in chief, 71, 75; proposes new ironclad bureau, 71; opinion of Isherwood and Lenthall, 71; removed from office, 72; testimony on light-draft monitors, 74–75
Stockton, Robert: 10
Strategy, naval: 144, 238, 249; influence of monitors on, 26, 248; influence of War of 1812 on, 32, 54; influence of ironclads of, 49; views of Isherwood and Lenthall on, 52–55, 64; based on coast defense and commerce destruction, 65; *see also* "Anaconda Policy"
Stuart, Charles B., civil and naval engineer: 15; as civil engineer, 8; appointed engineer in chief in 1850, 13; reliance on Isherwood, 13
Sumner, Senator Charles, opposes Isherwood's 1861 nomination: 4
Supercruisers: see *Wampanoag*
*Swatara*: 162

*Tacony*: 39
Tatham, G. N.: 214–15
Tatnall, Josiah, naval officer: 11, 12
Taylor model basin: see David W. Taylor Model Basin
Thurston, Robert Henry: 110; evaluation of Isherwood's experiments, 90, 96
*Ticonderoga*: 138–39
*Tonawanda*: 51
Tool purchases: see Zeller tool purchases
Toucey, Isaac: 106
*Tunxis*: 77
Turrets: 49, 51, 52, 55, 56, 57, 58, 63, 64, 66; Coles's, 50, 51, 56; Ericsson's, 50, 51; *see also* ironclads, turreted; monitors

*Uncle Samuel's Whistle* (pamphlet): 115

Van Horn, Representative Robert T.: 202, 223, 224, 225
*Virginia* (ironclad): 24, 26, 49, 52, 56, 59, 220, 221, 240; *see also Monitor*

Wade, Senator Benjamin F.: 198, 226
Wages, wartime: 38, 39, 40, 41, 42, 118
Walker, John G., naval officer: 206–207, 225
*Wampanoag*: 32, 145, 148, 168, 201, 202, 213, 220, 224, 238; need for such a class of warships, 169–70; dimensions, 170; hull design, 170–71; construction materials of, 170, 171, 172; machinery of, 171–72, 174; designed as a ram, 172; rigging of, 172; armament and ammunition, 172–73; construction of, 173–74; 259; launched in 1864, 173–74; cost of, 173, 177, 234; criticisms of, 174–75, 176, 177; sea trials, 177, 178–80; speed of, 178, 184; speed compared with merchant steamers and other warships, 179–80; engineering crew on final trial, 178; engine efficiency, 179; steaming endurance, 179; potential as a commerce destroyer, 181; forces British Navy to produce *Inconstant*, 183, 188, 261; line-officer evaluation of, in 1868, 183–85; alterations recommended, 184–85; decommissioned, 185; influence of on foreign interventions in Civil War, 187; influence of on *Alabama* claims, 187–88; criticized by Porter, 205; overhauled and made ineffective as a warship, 234; name changed to *Florida*, 234; sold in 1885, 234; derivation of name, 256; crew at commissioning, 259–60;

novel design features, 260; honored by Navy Department in 1930's, 260; Isherwood's opinion of trial, 260; speed trial criticized by Church, 260; influence on later warship design, 261
War Department: 135
War of 1812: and American naval strategy, 32; lessons from, 54
Warner, J. H., naval engineer: 107–108
*Warrior* (British): 49–50
Warships: naming of, 256–57; speed of, *see* speed
Washburne, Representative Elihu B.: 181, 224
Waterman, Henry: 110
*Water Witch:* 15
Weale, John: 8
Webb, William S.: 63–64
Weed, Thurlow: 136
Welles, Gideon: and Isherwood's 1861 nomination, 3–5, 21; and Gosport Navy Yard crisis, 21–22, 25, 26; and blockade of the Confederacy, 27; remarks about Lenthall, 28; and ninety-day gunboats, 30–31; and inexperienced naval engineers, 34; and need for government manufacturing facilities, 41; on vital role of steam Navy, 43–44; supports Isherwood's and Lenthall's seapower views, 55; indecision over ironclad policy, 61–62; decision not to support Isherwood's and Lenthall's ironclad program, 65; investigation of light-draft-monitor failure, 72–74; views on Fox and light-draft-monitor episode, 75–76; opinion of Stimers, 75–76; opinion of Dickerson, 114, 121, 124; criticizes New York city press, 124, 135–36; opposes Isherwood's ideas on steamer design, 167–78; views on Board of Admiralty plan, 206; opinion of Porter, 206, 207; on Porter and Board of Admiralty, 225; evaluation of Isherwood, 231–32
Whitney, William C.: 240
Wilkes, Charles, naval officer: 198
*Winooski:* 254; description of, 119, 126; dock trial and sea trial procedure, 123; engine performance of, 129; *see also Algonquin*
Wise, Henry A., naval officer: 210
Wood, William W. W., naval engineer: 215
Worden, John L., naval officer: 201
Wright, William, marine engineer: 123

Yulee, Senator, D. L.: 106

Zeller, Theodore: 241; on *Michigan* experiment board, 19–20; description of, 20; friendship with Isherwood, 20; on *Wampanoag* trial, 178–79; purchase of machine tools, 213–18; with Isherwood in New York on experiment boards, 235; European tour, 1875–76, with Isherwood, 236; work with Isherwood in retirement, 240; political affiliation, 264
Zeller tool purchases: 213–19, 224, 228, 230; Kelley's subcommittee investigation of, 214, 215–17, 264; House Naval Affairs committee investigation of, 218
"Zeller Tool Scandal": *see* Zeller tool purchases

*Edward William Sloan, III,* who holds the B.A. degree from Yale University, the M.A. degree from Yale University and Harvard University, and the Ph.D. degree from Harvard, is assistant professor of history in Trinity College, Hartford, Connecticut. He is also consultant in naval science and history to the New Campuses Program, University of California and visiting lecturer at the Munson Institute of American Maritime History, Mystic Seaport.

*Benjamin Franklin Isherwood, Naval Engineer: The Years as Engineer in Chief, 1861–1869* was set on the Linotype in eleven point Old Style No. 1, with two points leading between lines. Old Style No. 1, a nineteenth-century modification by the American foundry of MacKellar, Smiths, and Jordan of earlier English type, was selected by designer Gerard Valerio for its simplicity and freedom from frills. For pleasant readability and suitableness for the type, Warren's Olde Style paper, sixty-pound weight, was used. The book was composed, printed, and bound by the George Banta Company, Incorporated, Menasha, Wisconsin.